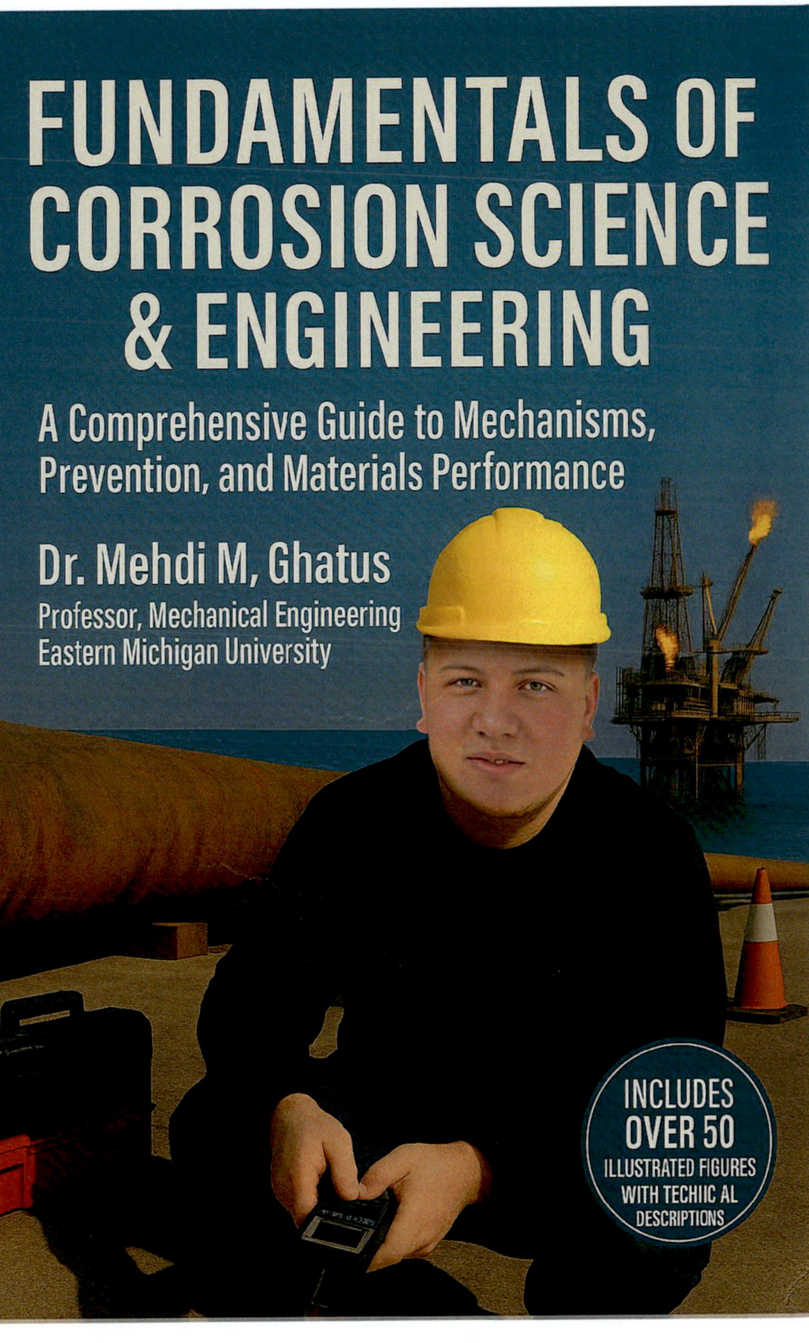
FUNDAMENTALS OF
CORROSION SCIENCE
& ENGINEERING
A Comprehensive Guide to Mechanisms,
Prevention, and Materials Performance
Dr. Mehdi M, Ghatus
Professor, Mechanical Engineering
Eastern Michigan University
INCLUDES
OVER 50
ILLUSTRATED FIGURES
WITH TECHIIC AL
DESCRIPTIONS

Fundamentals of Corrosion Science and Engineering

Mehdi Mohammed Ghatus

Fundamentals of Corrosion Science and Engineering

Mehdi Mohammed Ghatus
GameAbove College of Engineering, Mechanical Engineering
Eastern Michigan University
Ypsilanti, MI, USA

ISBN 978-3-032-13137-9 ISBN 978-3-032-13138-6 (eBook)
https://doi.org/10.1007/978-3-032-13138-6

This Springer imprint is published by the registered company Springer Nature Switzerland AG
The registered company address is: Gewerbestrasse 11, 6330 Cham, Switzerland

Dedication

I dedicate this humble work to my dear mother’s soul & the soul of my cousin, a teacher and a friend
Fatima Salem Alameen
Nasreddin Bunni

الاهداء

:المتواضع أهدي هذا العمل الى
روح امي الغالية فاطمه سالم الامين
روح ابن عمتي وأستاذي وصديقي نصرالدين بوني

English Language Reviewer

Professor William Anderson is an Emeritus Professor of Aerospace Engineering at The University of Michigan, USA. He has published more than 50 papers and has had 18 doctoral students. His research has concentrated on aeroelasticity and optimization. He volunteered to help the author in proofreading this book for English errors and has helped it read more smoothly. This is a good example of international collaboration.

Acknowledgment

The completion of this book would not have been possible without the participation and support of many individuals, whose names may not all be mentioned here. Their contributions are deeply appreciated and sincerely acknowledged. However, I would like to express my heartfelt gratitude in particular to the following:

Professor William Anderson is an Emeritus Professor of Aerospace Engineering at the University of Michigan, USA. With more than 50 published papers and the supervision of 18 doctoral students, his research has focused on aeroelasticity and optimization. Professor Anderson generously volunteered to proofread this book, significantly enhancing its clarity and readability. As he remarked, "This is a good example of international collaboration."

My wife, Samira Shaban Ghatus, for her unwavering support—both spoken and unspoken—throughout this journey.

Contents

About the Author

Mehdi Mohamed Ghatus is an international expert in corrosion science and mechanical engineering with a career spanning academia, research, and industry. He holds a Doctor of Engineering and M.S. in Mechanical Engineering from Lawrence Technological University (USA), a Master's degree from the University of Manchester in corrosion science (UK), and a Bachelor's degree from the University of Tripoli (Libya).

Dr. Ghatus began his professional journey in corrosion engineering at both onshore and offshore facilities, serving as Head of Corrosion Engineering at the Bouri Oil Field (Agip Oil and Gas Company). His expertise has been shaped by consulting roles with Mellitah Oil and Gas (Libya), Cooper Standard Automotive (USA), and teaching appointments at universities in Libya, China, and the United States.

He has published extensively in materials science, metallurgy, and mechanical engineering and currently serves as a visiting faculty member in China and as an academic delegate from Eastern Michigan University. His work bridges rigorous scientific research with hands-on industrial applications, offering deep insight into the principles and practices of corrosion protection.

Eastern Michigan University, Ypsilanti, MI, USA
Mehdi_gat@yahoo.com
https://www.emich.edu/cet/faculty/m-ghatus.php

List of Figures

List of Tables

Chapter 1
Introduction

USEFUL CONVERSION FACTORS AND RELATIONSHIPS

Length

SI unit: meter (m)

1 km = 0.62137 mi

1 mi = 5280 ft

= 1.6093 km

1 m = 1.0936 yd

1 in. = 2.54 cm (exactly)

1 cm = 0.39370 in.

1 Å = 10^{-10} m

Mass

SI unit: kilogram (kg)

1 kg = 2.2046 lb

1 lb = 453.59 g

= 16 oz

1 amu = 1.6605402 x 10^{-24} g

Temperature

SI unit: Kelvin (K)

0 K = −273.15°C

= −459.67°F

K = °C + 273.15

$°C = \frac{5}{9}(°F - 32°)$

$°F = \frac{9}{5}°C + 32°$

Energy (derived)

SI unit: Joule (J)

1 J = 1 kg-m^2/s^2

1 J = 0.2390 cal

= 1 C x 1 V

1 cal = 4.184 J

1 eV = 1.602×10^{-19} J

Pressure (derived)

SI unit: Pascal (Pa)

1 Pa = 1 N/m^2

= 1 kg/m-s^2

1 atm = 101,325 Pa

= 760 torr

= 14.70 lb/in^2

1 bar = 10^5 Pa

Volume (derived)

SI unit: cubic meter (m^3)

1 L = 10^{-3} m^3

= 1 dm^3

= 10^3 cm^3

= 1.0567 qt

1 gal = 4 qt

= 3.7854 L

1 cm^3 = 1 mL

1 in^3 = 16.4 cm^3

M. M. Ghatus, *Fundamentals of Corrosion Science and Engineering*, https://doi.org/10.1007/978-3-032-13138-6_1

Corrosion must be controlled in engineering materials, leading to significant losses in production, valuable resources, and even risks to human safety. While corrosion cannot be completely prevented, it can be effectively minimized. The two primary consequences of corrosion are the loss of valuable resources and product contamination. Corrosion also has a direct economic impact. According to recent estimates, corrosion accounts for approximately 3.4% of the global Gross Domestic Product (GDP), amounting to an international cost of around $2.5 trillion annually. Implementing effective corrosion control measures could reduce these costs by 15–35%, translating to annual savings between $375 billion and $875 billion.

Corrosion Definitions

1. The most common definition of corrosion is the deterioration in performance of an engineering system. This deterioration can lead to:

   - Loss of service life
   - Loss of efficiency
   - Loss of appearance
   - Loss of mechanical strength
   - Loss of product quality
   - Increased safety risks
   - Increased operational costs

   Corrosion is also understood as the tendency of a material to revert to its native state. For example, in the case of iron, its natural form is iron oxide, such as FeO, Fe_2O_3, or Fe_3O_4. During production, iron is obtained through the reduction of these oxides—essentially the removal of oxygen. When corrosion occurs, iron reacts with oxygen to form iron oxides. This cyclic transformation is illustrated in Fig. 1.1.

2. Electrochemical degradation of a metal occurs through its reaction with the surrounding environment, for example, the corrosion of zinc in a hydrochloric acid solution

$$Zn + 2Hcl \rightarrow Zncl_2 + H_2 \uparrow$$

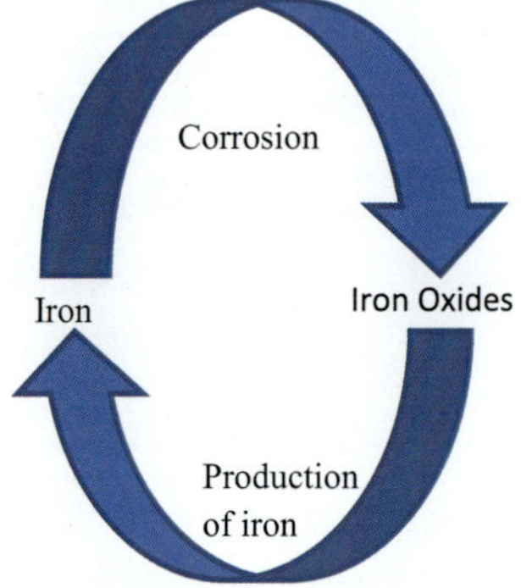

Fig. 1.1 Corrosion and production of iron

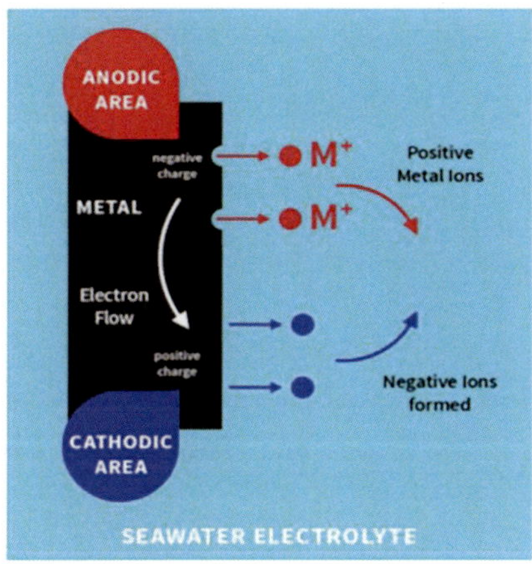

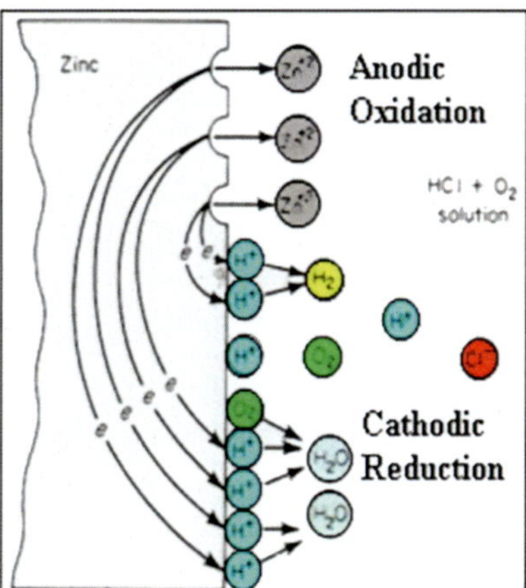

Fig. 1.2 Mechanisms of corrosion

This reaction happens in two stages:

- Oxidation process: $Zn - 2e \rightarrow Zn^{++}$.
- Reduction process: $2H^{+} + 2e \rightarrow 2H \rightarrow H_2$

The anodic and cathodic areas and migration of electrons are illustrated in Fig. 1.2.

For any corrosion process to occur, four essential components must be present:

- Anode, the area where metal is oxidized and corrosion occurs.
- Cathode, the area protected from corrosion where reduction reactions take place.
- Electrolyte, a conductive medium that allows ionic movement between the anode and cathode.
- Electrical conductor, a path for electron flow between the anode and cathode.

Chapter 2
Electrochemistry of Corrosion

Electrochemistry is the study of the relationship between chemical reactions and electrical energy. Certain chemical reactions can generate electricity. For example, when metal A (more active) is connected to metal B (less active) through an electrolyte or an external wire, electrons flow from metal A to metal B, as illustrated in Fig. 2.1. In this process, metal A undergoes corrosion (acts as the anode), while metal B is protected (acts as the cathode). This principle forms the basis of cathodic protection, a widely used technique to prevent metal corrosion.

2.1 Current Change Density

The exchange current density is the current that exists without net electrolysis and at zero overpotential. It can also be defined as the current density that flows equally in both the anodic and cathodic directions under equilibrium conditions. A higher exchange current density indicates a faster reaction rate, and vice versa.

In electrochemical systems, a metal can either oxidize, converting into metal ions (anodic reaction), or metal ions can gain electrons and deposit back as metal atoms (cathodic reaction). The oxidation process is referred to as anodic, while the reduction process is cathodic. These two simultaneous processes are illustrated by the arrows in Fig. 2.2.

$$M \rightleftarrows M^{+n} + ne$$

- The anodic reaction generates an anodic current (i_a), where metal atoms are oxidized into metal ions:

$$\mathrm{Metal} \rightarrow \mathrm{Metal}^{n+} + \mathrm{ne}^{-}$$

M. M. Ghatus, *Fundamentals of Corrosion Science and Engineering*,
https://doi.org/10.1007/978-3-032-13138-6_2

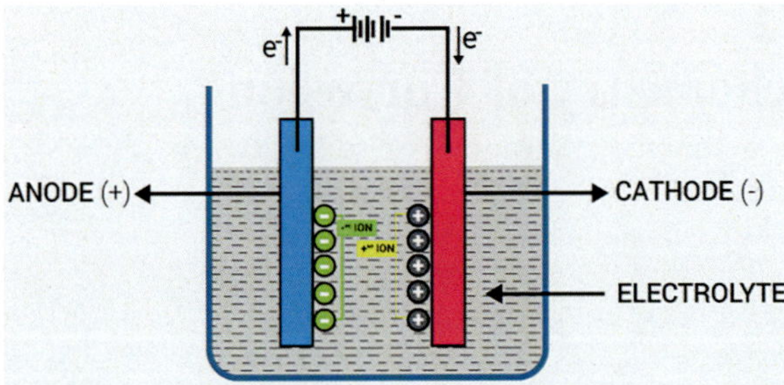

Fig. 2.1 Corrosion cell

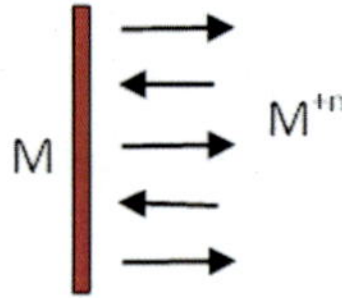

Fig. 2.2 Anodic and cathodic reactions

- The cathodic reaction generates a cathodic current (i_c), where metal ions are reduced and deposited as metal atoms:

$$\text{Metal}^{n+} + ne^{-} \rightarrow \text{Metal}$$

In equilibrium conditions, $i_a = i_c$ OR $i_a = -i_c$ (*negative sign means opposite direction only*)

$$\frac{i_a a_m}{nF} = \frac{i_c a_m}{nF} = i_a = i_c$$

where:

a_m is the atomic weight
n is the number of electrons
F is the Faraday constant

2.1.1 Factors Affecting Exchange Current Density

Several factors affect the exchange current density, including:

1. **Electrode material**, In a hydrogen evolution reaction, platinum has an exchange current density of approximately 10^{-2} to 10^{-3} A/cm^2, while zinc has a much lower value of about 10^{-10} to 10^{-11} A/cm^2.
2. **Surface condition**, Smoother, cleaner surfaces tend to enhance electrochemical activity.

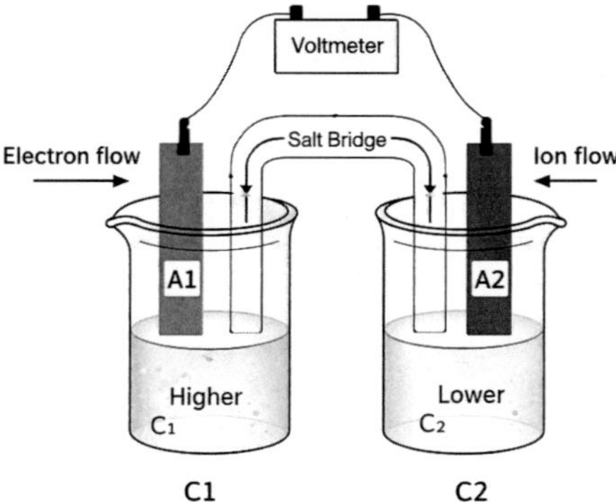

Fig. 2.3 Concentration difference

3. **Solution composition**, The type and concentration of ions in the electrolyte influence the reaction rate.
4. **Environmental conditions**, pH, agitation, and dissolved gases all play a role.
5. **Temperature**, Higher temperatures generally increase reaction rates and current densities.

As shown in Fig. 2.3, the concentrations C_1 and C_2 are not equal ($C_1 \neq C_2$); specifically, $C_2 > C_1$. As a result, the current flows from the region of higher concentration (C_2) to the region of lower concentration (C_1).

In addition, when two metals are electrically connected and the surface area of the anode (Aa) is smaller than that of the cathode (Ac), that is, Aa < Ac, this leads to a higher anodic current (ia > ic). This imbalance can significantly accelerate localized corrosion at the anodic site

$$\frac{i_a}{A_a} \neq \frac{i_c}{A_c} = i_a > -i_c$$

2.2 Corrosion Rate

The corrosion rate refers to the amount of metal lost due to corrosion per unit time. It can be calculated for both metals and alloys using current density and material-specific data. To accurately determine the corrosion rate, the following parameters must be obtained:

- Weight loss, the reduction in the metal's mass over a specified period
- Density, the density of the metal or alloy being evaluated
- Area, the total initial surface area of the exposed metal
- Time, the duration of the exposure or reference period

For example:

- mpy (milli – inch/year)

$$\text{mpy}\left(\frac{\text{milli}-\text{inch}}{\text{year}}\right)=\frac{i_o a}{nF\rho}$$

where:

- i_o = exchange current density (A/cm^2)
- a = atomic weight of the metal (g/mol)
- n = number of electrons involved in the electrochemical reaction
- F = Faraday's constant = 96,500 C/mol
- ρ = density of the metal (g/cm^3)
- mm/year (millimeter/year)

$$\frac{\text{mm}}{\text{year}}=87.6\left(\frac{w}{DAT}\right)$$

where:

- W = weight loss (mg)
- D = density of the metal (g/cm^3)
- A = surface area of the sample (cm^2)
- T = exposure time (hours) *ipy* (inch/year)

$$ipy=\frac{12w}{tA\rho}$$

where:

- w = mass loss over time t, in pounds (lb)
- t = time of exposure, in years (yr)A is the surface area, ft^2
- ρ is the density of the material lb/ft^3

$$mpy=\frac{534w}{DAT}$$

where:

- W = weight loss of the specimen (mg)
- D = density of the material (g/cm^3)
- A = surface area of the specimen exposed to corrosion (cm^2)
- T = duration of exposure (hours)
- Based on ASTM standard G1-90 (Reapproved 1999)

$$mpy = \frac{kw}{DAT}$$

- K = constant (unit-dependent; see Table 2.1)
- T = time of exposure (hours)
- A = surface area of the specimen (cm^2)
- W = mass loss (grams)

The corrosion rate can be calculated in a variety of units with the following appropriate value of K.

These constants can be used to convert corrosion rates between different unit systems. To convert a corrosion rate from units X to units Y, multiply the original rate by the ratio K_Y/K_X.

Example

$$15\,\text{mpy} = 15^{*}\left[\frac{\left(2.78^{*}\,10^{6}\right)}{\left(3.45^{*}\,10^{6}\right)}\right]\text{pm / s.} = 12.086\ \text{pm / s}$$

2.2.1 Calculation of Corrosion Rate

Metals

Example

Calculate the corrosion rate in mpy for iron (Fe); if the current change density of iron (Fe) is $1.69*10^{-4}$ A/cm^2, atomic weight of iron 55.85, and density (ρ) = 7.86 gm/cm^3.

Table 2.1 Values of constant (K) with different corrosion rate expressions

Corrosion Rate Units Desired	Constant (K) in corrosion rate equation
Mils per year (mpy)	3.45×10^6
Inches per year (ipy)	3.45×10^3
Inches per month (ipm)	2.87×10^2
Millimeter per year (mm/y)	8.76×10^4
Micrometer per year (μm/y)	8.76×10^7
Picometers per second (pm/s)	2.78×10^6
Grams per square meter per hour (g/m^2.h)	1.00×10^4x D
Milligram per square decimeter per day (mdd)	2.40×10^6x D
Microgram per square meter per second (μg/m^2.s)	2.78×10^6x D

Solution

$$\text{mpy} = \frac{i_o a_{Fe}}{nF\rho_{Fe}}$$

$$\text{mpy} = \frac{1.69 * 10^{-4} \frac{\text{A}}{\text{cm}^2} 55.85\,\text{gm}}{2 * 96500 * 7.86 \frac{\text{gm}}{\text{cm}^3}}$$

$$\text{mpy} = \frac{1.69 * 10^{-4} \frac{\frac{\text{culomb}}{\text{cm}^2}}{\text{sec}} 55.85\,\text{gm}}{2 * 96500 * 7.86 \frac{\text{gm}}{\text{cm}^3}}$$

$$\text{mpy} = \frac{1.69 * 10^{-4}\, 55.85 * 10^3 * 24 * 365 * 60 * 60}{2 * 96500 * 7.86 * 2.54}$$

$$\text{mpy} = 77.251 \frac{\text{milli} - \text{inch}}{\text{year}} = 77.251 \frac{\text{mils}}{\text{year}}$$

Alloys

When two or more metals are combined, the resulting material is called an alloy. The total composition of all elements in the alloy must equal 100%. For example, if an alloy is composed of metals A, B, and C with respective percentages x, y, and z, then:

$$\mathbf{x\% + y\% + z\% = 100\%}$$

In other words, the overall composition of the alloy can be represented as:

$$\mathbf{A_x + B_y + Cz}$$

To calculate the corrosion rate of an alloy using current density, the following parameters must first be determined:

$$[a/n]_{Equivalent} = \sum f_j M_j / n_j$$

where:

j = jth element =1, 2, 3, …, etc. (Metals in the alloy)
f = fraction (%)
M = metal
n = valence (number of electrons)

$$\rho_{alloy} = \sum f_j \rho_j$$

Example
Calculate the corrosion rate of AISI 316 alloy at $i = 1\ \mu A/cm^2$.

Metal	Weight %	Valence (n)	Density (ρ) gm/cm³	Atomic wt (a) gm/mol
Cr	18	1	7.1	52.01
Ni	8	2	8.9	58.68
Mo	3	1	10.2	95.95
Fe	70	2	7.86	55.85

Solution

$$[a/n]_{Equivalent} = \sum f_j M_j / n_j = \left(\frac{0.18^*52.01}{1} + \frac{0.08^*58.68}{2} + \frac{0.03^*95.95}{1} + \frac{0.7^*55.85}{2} \right) = 35.135 \text{ gm}$$

$$\rho_{alloy} = \sum f_j \rho_j = 0.18^*7.1 + 0.08^*8.9 + 0.03^*10.2 + 0.7^*7.86 = 7.782 \text{ gm/cm}^3$$

$$\text{mpy} = \frac{i_o \left(\frac{a}{n} \right)_{equivalent}}{F \rho_{alloy}} = \frac{1 \times 35.135}{96500 \times 7.782} = 0.56 \text{mpy}$$

Example
Corrosion of a galvanized steel sheet with a zinc coating of 0.1 mm thick generates a corrosion current of 6×10^{-3} A/m². Determine if the coating is sufficient to give rust-free protection for 10 years (Density of zinc = 7.13 mg/m³, atomic weight of zinc = 65.4).

Solution

$$Zn \rightarrow Zn^{++} + 2e$$

$$\text{Current generated / m}^2 = 6 \times 10^{-3} \text{A} = 6 \times 10^{-3} \text{c/s}$$

For 10 years, the charge generated = $6 \times 10^{-3} \times 3600 \times 24 \times 365 \times 10$

Table 2.2 Corrosion rate, unit converter

Corrosion Rate Unit Converter								
CorrRateUnitConverter converts between all corrosion	From	µA/cm^2	To	mpy	µm/y	mm/y	gmd	mdd
rate units for all metals and alloys.		1.0000	=	0.4134	10.5002	0.0105	0.2267	2.2670
µA/cm^2: micro-ampere per cm^2	From	mdd	To	mpy	µm/y	mm/y	gmd	µA/cm^2
mpy: milli-inch per year		1.0000	=	0.1824	4.6317	0.0046	0.1000	0.4411
µm/y: micrometer per year mm/y: millimeter per year	From	gmd	To	mpy	µm/y	mm/y	µA/cm^2	mdd
		1.0000	=	1.8235	46.3172	0.0463	4.4111	10.0000
gmd: gram per m^2 per day	From	µm/y	To	mpy	µA/cm^2	mm/y	gmd	mdd
mdd: milligram per dm^2 per day		1.0000	=	0.0394	0.0952	0.0010	0.0216	0.2159
	From	mpy	To	µA/cm^2	µm/y	mm/y	gmd	mdd
		1.0000	=	2.4190	25.4000	0.0254	0.5484	5.4839

$$\text{Number of electrons produced in 10 years} = \frac{6\times10^{-3}\times3600\times24\times365\times10}{1.6\times10^{-19}} = \frac{1.18\times10^{25}\,\text{es}}{\text{m}^2}$$

$$\text{Number of atoms} = \frac{\dfrac{1.18\times10^{25}\,\text{es}}{\text{m}^2}}{2} = 5.89\times10^{24}\ \text{atoms}$$

$$\text{Masss of zinc} = \frac{5.89\times10^{24}}{6.022\times10^{23}}\times65.4 = 640\,\text{g} = 0.64\,\frac{\text{kg}}{\text{m}^2}$$

$$\begin{aligned}\text{Density} &= 7.13\times10^{6}\ \ \text{g/m}^3\\ &= 7.13\times10^{3}\ \ \text{kg/m}^3\end{aligned}$$

$$\text{Thickness} = \frac{0.64\,\text{kg/m}^2}{7.13\times10^{3}\ \text{kg/m}^3} = 9\times10^{-5}\ \text{m} = 0.09\,\text{mm}$$

To protect iron (Fe) from corrosion for 10 years, a zinc coating thickness of 0.09 mm is required. Therefore, a 0.1 mm zinc layer is generally considered sufficient to provide 10 years of protection under standard environmental conditions (Table 2.2).

Chapter 3
Thermodynamics of Corrosion, Electrochemical Series, and Concentration Cell

Free energy (G) represents the energy available to perform work under conditions of constant temperature and pressure. In thermodynamics, Gibbs free energy determines whether a chemical reaction will occur spontaneously.

- If $\Delta G > 0$, the reaction is nonspontaneous.
- If $\Delta G < 0$, the reaction is spontaneous.
- If $\Delta G = 0$, the system is at equilibrium and no net reaction occurs.
- If the reaction;

$$\mathrm{aA + bB \leftrightarrow cC + dD}$$

where a, b, c, and d are the number of moles. A and B are the reactants, and C and B are the products.

In the forward reaction, the rate of the reaction (R) is equal to:

$$\vec{R} = k_1 C'^{a}_{A} * C'^{b}_{B}$$

where C'_A is the concentration of substance A, as well as C'_B to B. $\mathbf{k_1}$ is the rate constant of the forward reaction.

In the backward reaction, the rate of the reaction (R) is equal to:

$$\overleftarrow{R} = k_2 C'^{c}_{C} * C'^{d}_{D}$$

where C'_C is the concentration of substance C, as well as C'_D to D. $\mathbf{k_2}$ is the rate constant of the backward reaction.

At equilibrium condition

M. M. Ghatus, *Fundamentals of Corrosion Science and Engineering*,
https://doi.org/10.1007/978-3-032-13138-6_3

$$\bar{R} = \vec{R} = \frac{K_1}{K_2} = \frac{C'^{c}_{C} * C'^{d}_{D}}{C'^{a}_{A} * C'^{b}_{B}} = K_C$$

where the K_c is the equilibrium constant, when the temperature remains constant, if the activity (a) is considered rather than the concentration, the last equation will be rewritten as follows:

$$\bar{R} = \vec{R} = \frac{K_1}{K_2} = \frac{a^{c}_{C} * a^{d}_{D}}{a^{a}_{A} * a^{b}_{B}} = K_a$$

where a = f C when the solution is very diluted f = 1, therefore, a = c (laws of mass action).

The free energy (G) is a function of pressure, temperature, and number of moles; therefore,

$$\mathrm{G} = \mathrm{f}\left(\mathrm{P,T,n_1,n_2 \ldots n_n}\right)$$

$$dG = \left(\frac{\partial G}{\partial P}\right)_{T,n1,n2\ldots n_n} dP + \left(\frac{\partial G}{\partial T}\right)_{P,n1,n2\ldots n_n} dT$$
$$+ \left(\frac{\partial G}{\partial n_1}\right)_{P,T,n2\ldots n_n} dn_1 + \ldots + \left(\frac{\partial G}{\partial n_n}\right)_{T,P,n1,n2\ldots n_{n-1}} dn_n$$

This is called a partial molal quantity, which is equal to chemical potential (μ) therefore,

$$dG = \mu_1 dn_1 + \mu_2 dn_2 + \ldots + \mu_n dn_n$$

where T and P are both constants.

If the system has several components (n) at constant temperature and pressure, and the change in each component is equal to Δx, then

$$\Delta x dG = \Delta x \mu_1 dn_1 + \Delta x \mu_2 dn_2 + \ldots + \Delta x \mu_n dn_n$$

$$G = \mu_1 n_1 + \mu_2 n_2 + \ldots + \mu_n dn_n$$

$$\Delta G = G_{product} - G_{reactant}$$

$$\mathrm{aA + bB \leftrightarrow cC + dD}$$

$$G_{reactant} = a\mu_a + b\mu_b$$

$$G_{product} = c\mu_c + d\mu_d$$

$$\mu = \mu^{\circ} + RTlna$$

- For the substance A
- $\mu_A = \mu_A^{\circ} + RTlna_A$
- For substance B
- $\mu_B = \mu_B^{\circ} + RTlna_B$
- For the substance C
- $\mu_C = \mu_C^{\circ} + RTlna_C$
- For the substance D
- $\mu_D = \mu_D^{\circ} + RTlna_D$
- $\Delta G = G_{product} - G_{reactant}$

$$c\mu_C^{\circ} + cRTlna_C + d\mu_D^{\circ} + dRTlna_D - a\mu_A^{\circ} - aRTlna_A - b\mu_B^{\circ} - bRTlna_B = \Delta G$$

$$c\mu_C^{\circ} + d\mu_D^{\circ} - a\mu_A^{\circ} - b\mu_B^{\circ} + RTln\frac{a_C^c * a_D^d}{a_A^a * a_B^b} = \Delta G$$

$$\Delta G = Z + RTln\frac{a_C^c * a_D^d}{a_A^a * a_B^b}$$

where Z is the constant

$$\Delta G = \Delta G^{\circ} + RTln\frac{a_C^c * a_D^d}{a_A^a * a_B^b}$$

At equilibrium ΔG = 0

$$\Delta G^{\circ} = -RTln\frac{a_C^c * a_D^d}{a_A^a * a_B^b}$$

$$\Delta G^{\circ} = -RTlnk_a$$

In the case of using the concentration

$$\Delta G^{\circ} = -RTlnk_c$$

In any corrosion reaction, both oxidation and reduction processes must occur. These electrochemical processes take place in one of two types of cells: the galvanic cell and the electrolytic cell. In a galvanic cell, chemical energy is spontaneously converted into electrical energy. In contrast, an electrolytic cell requires an external electrical current to drive a nonspontaneous chemical reaction.

As shown in Fig. 3.1, when copper and zinc are connected and immersed in copper sulfate and zinc sulfate solutions, respectively, the zinc (being more active) acts as the anode, while the copper (less active) acts as the cathode.

The anodic and cathodic reactions of zinc

$$Zn - 2e \leftrightarrows Zn^{++} \quad \mathrm{E}^{\circ} = -0.76 \ \mathrm{V}$$

The anodic and cathodic reactions of copper

$$Cu - 2e \leftrightarrows Cu^{++} \quad \mathrm{E}^{\circ} = 0.34 \ \mathrm{V}$$

The voltmeter reading will be 0.34 – (−0.76) = 1.1v.

The potential difference in this system arises from redox reactions, involving both anodic and cathodic processes. Specifically, the copper ion has a greater tendency to be reduced and deposited as metallic copper than the zinc ion. Through this redox reaction, chemical energy is converted into electrical energy.

The rate of the reaction depends on the activation energy required to initiate the process. As illustrated in Fig. 3.2, the free energy gradient (ΔG) is negative, indicating that the reaction is spontaneous. In other words, the energy available to perform useful work in this system appears as electrical energy.

For example, the electric charge (Q) required to deposit one gram mole of copper on the electrode is one Faraday (F), equal to 96,500 coulombs per gram mole. This amount of charge is necessary to reduce one mole of copper ions (Cu^{2+}) to one mole of metallic copper (Cu^0). The total Q = nF, where n is the number of moles

$$\Delta G = \mathrm{EQ}$$
$$\Delta G = -\mathrm{nFE}$$
$$\Delta G^{\circ} = -\mathrm{nFE}^{\circ}$$

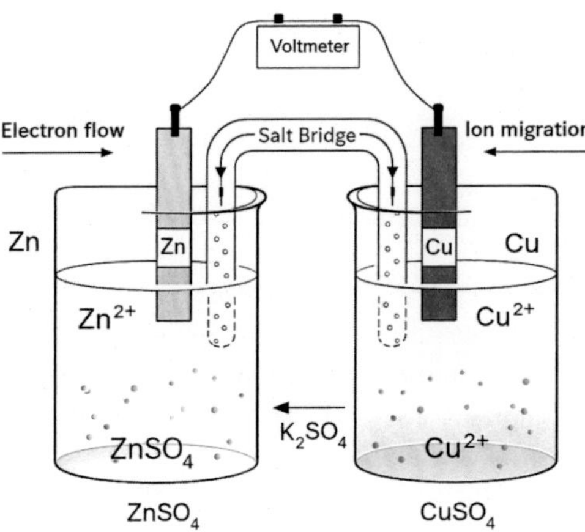

Fig. 3.1 Copper and zinc are connected and immersed in copper sulfate and zinc sulfate

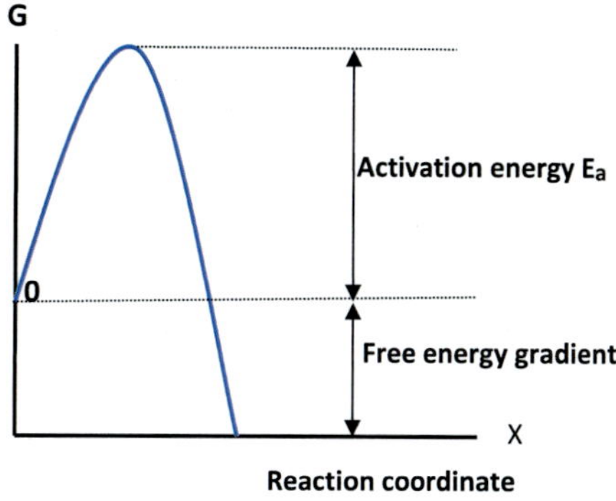

Fig. 3.2 Free energy gradient

$$Z^{++} + 2e \rightarrow Zn$$

$$E_{\frac{Zn^{++}}{Zn}} = E^{\circ}_{\frac{Zn^{++}}{Zn}} + \frac{RT}{2F} \ln \frac{[Zn^{++}]}{[Zn]} \quad \text{Reduction potential}$$

$$Zn \rightarrow Z^{++} + 2e$$

$$E_{\frac{Zn}{Zn^{++}}} = E^{\circ}_{\frac{Zn}{Zn^{++}}} + \frac{RT}{2F} \ln \frac{[Zn]}{[Zn^{++}]} \quad \text{Oxidation potential}$$

If:

$$Zn \rightarrow Z^{++} + 2e$$

$$Cu^{++} + 2e \rightarrow Cu$$

The overall redox is:

$$Zn + Cu^{++} \rightarrow Zn^{++} + Cu$$

In this reaction:

- Zinc is oxidized to Zn^{2+} (loses electrons at the anode).
- Copper ions (Cu^{2+}) are reduced to metallic copper (gain electrons at the cathode)
- $$E = E^{\circ} + \frac{RT}{2F} \ln \frac{[Zn][Cu^{++}]}{[Zn^{++}][Cu]}$$
- $$E = E^{\circ} + \frac{RT}{2F} \ln \frac{1}{k_c}$$

$$\Delta E^\circ = E^\circ_{\frac{Cu^{++}}{Cu}} - E^\circ_{\frac{Zn^{++}}{Zn}}$$

$$\Delta E^\circ = 0.34 - (-0.76) = 1.1\ V$$

$$\Delta G = -nFE$$

$$\Delta G^\circ = -nFE^\circ$$

$$\frac{\Delta G}{-nF} = \frac{\Delta G^\circ}{-nF} + \frac{RT}{nF}\ln\frac{[Zn]}{[Zn^{++}]}$$

$$\Delta G = \Delta G^\circ - RTln\frac{[Zn]}{[Zn^{++}]}$$

For the overall reaction (Fig. 3.3):

$$\Delta G = \Delta G^\circ - RTln\frac{[Zn^{++}][Cu]}{[Cu^{++}][Zn]}$$

$$\Delta G = \Delta G^\circ - RTlnk_c$$

$$\Delta E = E_2 - E_1$$

where the reduction potential of $\mathbf{E_2 > E_1}$ (Table 3.1).

For quantification, some standards are required:

- Hydrogen electrode, the notation for hydrogen electrode is **H_2 (1atm) {Pt} |$H^+_{(a=1)}$**. The schematic of the hydrogen electrode is shown in Fig. 3.4.

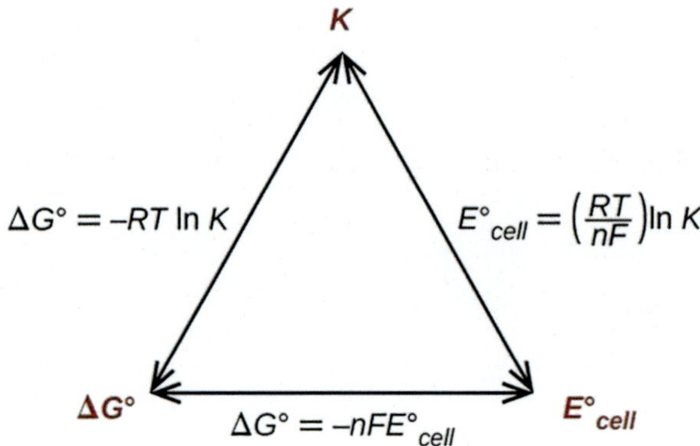

Fig. 3.3 Relationships between K, E°, and ΔG°

Table 3.1 Concludes the relationship between ΔG°, K, and E°

ΔG°	K	E°
Negative	>1	Positive
Zero	1	Zero
Positive	<1	Negative

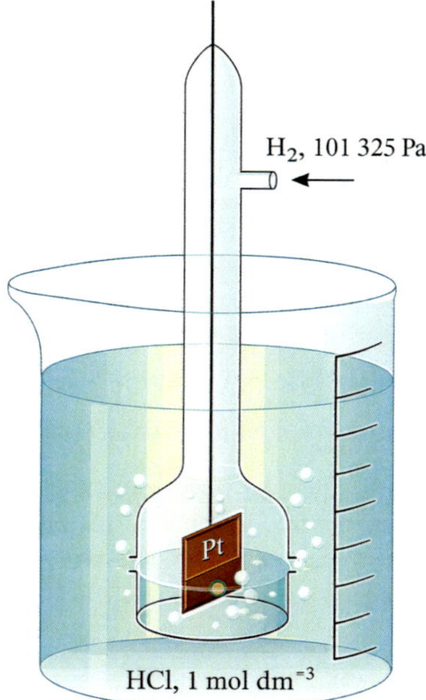

Fig. 3.4 Standard Hydrogen Electrode (SHE)

- $$2H^+ + 2e \leftrightarrows 2H_2$$
- $$E_{\frac{H^+}{\frac{1}{2}H_2}} = E^\circ_{\frac{H^+}{\frac{1}{2}H_2}} + \frac{RT}{1F}\ln\frac{\left[H^+\right]}{\left[P_{H_2}\right]}$$

$\mathbf{P_{H2}}$ is used because the final product is a gas, the activity = fC a = 1

$$E_{\frac{H^+}{\frac{1}{2}H_2}} = E^\circ_{\frac{H^+}{\frac{1}{2}H_2}} + 0$$

$$E^\circ_{\frac{H^+}{\frac{1}{2}H_2}} = 0$$

$$E_{\frac{H^+}{\frac{1}{2}H_2}} = 0 \quad \text{Standard hydrogen electrode}(\text{SHE})$$

For example, when copper is immersed in a hydrochloric acid (HCl) solution, its electrode potential relative to the Standard Hydrogen Electrode (SHE) can be measured. This is done by electrically connecting the copper electrode to the SHE, allowing for direct comparison of their potentials within the electrochemical cell.

$$\Delta E^\circ = E^\circ_{\frac{Cu^{++}}{Cu}} - E^\circ_{\frac{H^+}{\frac{1}{2}H_2}}$$

$$0.34 = E^\circ_{\frac{Cu^{++}}{Cu}} - 0$$

$$E^\circ_{\frac{Cu^{++}}{Cu}} = 0.34\,V$$

Standard Electrode Potentials

The standard reduction potentials of common metals, measured on the hydrogen scale at 298 K, are presented in Table 3.2. These values are essential in understanding the electrochemical behavior of metals under standard conditions.

Advantages of Standard Electrode Potentials

- Distinguish between active metals and noble metals.
- Help predict galvanic corrosion when two metals are electrically connected:
- The more active metal will corrode (sacrifice itself) to protect the more noble metal.

Table 3.2 Standard reduction potential at 298°K on the hydrogen scale

Half Reaction				Standard Potential (V)
F_2	$+ 2e^-$	$\rightleftharpoons$	$2F^-$	+2.87
Pb^{4+}	$+ 2e^-$	$\rightleftharpoons$	Pb^{2+}	+1.67
Cl_2	$+ 2e^-$	$\rightleftharpoons$	$2Cl^-$	+1.36
$O_2 +$	$4H^+ + 4e^-$	$\rightleftharpoons$	$2H_2O$	+1.23
Ag^+	$+ e^-$	$\rightleftharpoons$	Ag	+0.80
Fe^{3+}	$+ e^-$	$\rightleftharpoons$	Fe^{2+}	+0.77
Cu^{2+}	$+ 2e^-$	$\rightleftharpoons$	Cu	+0.34
$2H^+$	$+ 2e^-$	$\rightleftharpoons$	H_2	0.00
Pb^{2+}	$+ 2e^-$	$\rightleftharpoons$	Pb	−0.13
Pb^{2+}	$+ 2e^-$	$\rightleftharpoons$	Pb	−0.13
Fe	$+ 2e^-$	$\rightleftharpoons$	Fe	−0.44
Zn^{2+}	$+ 2e^-$	$\rightleftharpoons$	Zn	−0.76
Al^{3+}	$+ 3e^-$	$\rightleftharpoons$	Al	−1.66
Mg^{2+}	$+ 2e^-$	$\rightleftharpoons$	**Mg**	−2.36
Zl	$+ e^-$	$\rightleftharpoons$	**Li**	−3.05
Li^+	$+ 1e^-$	$\rightleftharpoons$	Zn	−2.76
Mg^{2+}	$+ 2e^-$	$\rightleftharpoons$	**Mg^{2+}**	−2.36
Al^{3+}	$+ 3e^-$	$\rightleftharpoons$	**Al**	−2.36
Mg^{2+}	$+ 2e^-$	$\rightleftharpoons$	**Mg**	−3.05
Li^+	$+ 1e^-$	$\rightleftharpoons$	**Li**	−3.05

stronger oxidizing agent ↑

stronger reducing agent ↓

stronger reducing agent ↓

Limitations of Standard Electrode Potentials

- Applicable only to pure metals, not alloys.
- Valid only at standard temperature (298 K).
- Provide no information about the rate of corrosion or metal deposition.

Example

A zinc electrode is immersed in a zinc chloride solution. If the activity of zinc in the solution is equal to (a_{Zn++}) 10^{-3}, what is the potential developed in this electrode?

Solution

$$E^{\circ}_{\frac{Zn^{++}}{Zn}} = -0.761\,V$$

$$E_{\frac{Zn^{++}}{Zn}} = E^{\circ}_{\frac{Zn^{++}}{Zn}} + \frac{RT}{2F}\ln\frac{\left[aZn^{++}\right]}{\left[aZn\right]}$$

$$E_{\frac{Zn^{++}}{Zn}} = -0.761 + \frac{8.134 * 298}{2 * 96500}\ln\frac{\left[10^{-3}\right]}{[1]}$$

$$E_{\frac{Zn^{++}}{Zn}} = -0.82\,v$$

Example: Corrosion Tendency of Nickel in Deaerated Water

Nickel is immersed in deaerated water (oxygen-free) with a pH of 8. The solubility product (K_{sp}) of $Ni(OH)_2$ is 1.6×10^{-16}, and the standard electrode potential of nickel is $E° = -0.25$ V (vs SHE).

Determine whether nickel tends to corrode under these conditions.

Solution

The anodic and the cathodic reactions of Ni in deaerated water are:

$$\mathrm{Ni} \rightarrow \mathrm{Ni}^{++} + 2\mathrm{e}\,(\text{anodic reaction})$$

$$2\mathrm{H}^{+} + 2\mathrm{e} \rightarrow \mathrm{H}_2\ \ (\text{cathodic reaction})$$

$$E_{\frac{Ni^{++}}{Ni}} = E^{\circ}_{\frac{Ni^{++}}{Ni}} + \frac{RT}{2F}\ln\frac{\left[aNi^{++}\right]}{\left[aNi\right] = 1}$$

$$E_{\frac{H^{+}}{\frac{1}{2}H_2}} = E^{\circ}_{\frac{H^{+}}{\frac{1}{2}H_2}} + \frac{RT}{2F}\ln\frac{\left[H^{+}\right]^2}{\left[P_{H_2}\right] = 1}$$

$$\Delta E = \mathrm{E}_{\text{cathode}} - \mathrm{E}_{\text{anode}}$$

$$\Delta E = \left(E_{\frac{H^{+}}{\frac{1}{2}H_2}} - E_{\frac{Ni^{++}}{Ni}}\right) + \frac{RT}{2F}\left[\ln\left[H^{+}\right]^2 - \ln Ni^{++}\right]$$

$$\Delta E = (0-(-0.25)) + \frac{RT}{2F}\left[2.303\log\left[H^+\right]^2 - 2.303\log Ni^{++}\right]$$

$$\Delta E = (0.25) + \frac{2.303RT}{2F}\left[-(-\log\left[H^+\right]^2 - \log Ni^{++}\right]$$

$$\Delta E = (0.25) + \frac{2.303RT}{2F}\left[-2pH - \log Ni^{++}\right]$$

$$\Delta E = (0.25) - \frac{2*2.303RT*pH}{2F} - \frac{2.303RT}{2F}\log Ni^{++} *****$$

$$Ni^{+2} + 2OH^- = Ni(OH)_2$$

$$H_2O = H^+ + OH^-$$

$$10^{-14} = \left[H^+\right]\left[OH^-\right]$$

$$\log\left[H^+\right] + \log\left[OH^-\right] = -14$$

$$-\log\left[H^+\right] - \log\left[OH^-\right] = 14$$

$$pH - \log\left[OH^-\right] = 14$$

$$\left[OH^-\right] = 10^{-6}$$

$$K_{sp} = \left[Ni^{+2}\right]\left[OH^-\right]^2$$

$$\left[Ni^{+2}\right] = \frac{1.6*10^{-16}}{\left[10^{-6}\right]^2} = 1.6*10^{-4}\ unites$$

Substitute in equation *****.

R = 8.314 J/mole/kelvin, Temperature 298°K, and F = 96,500 coulomb/gr equivalent

$$\Delta E = (0.25) - \frac{2*2.303*8.314*298*8}{2*96500} - \frac{2.303*8.314*298}{2*96500}\log 1.6*10^{-4}\ \Delta E = -0.11v\Delta G = -nFE = -2*96500(-0.11) = 21380.2(Positive).$$

Conclusion

Nickel will not corrode in deaerated water at pH 8 because the system does not support a spontaneous redox reaction under these conditions. The absence of dissolved oxygen prevents a cathodic reaction, and the calculated electrode potential suggests a nonspontaneous process. Since ΔG > 0, the reaction is not thermodynamically favorable, and thus, corrosion will not occur.

Example

An electrochemical cell consists of pure copper and lead electrodes, each immersed in a solution containing their respective divalent ions. The concentration of Cu^{2+} is 0.6 M, and the observed cell potential is 0.507 V at 25 °C (298 K).

Assuming standard conditions otherwise, calculate the concentration of Pb^{2+} ions.

Solution

$$Cu^{2+} + 2e \leftrightarrows Cu$$

$$Pb \leftrightarrows 2e + Pb^{2+}$$

$$\Delta E^{\circ} = E^{\circ}_{oxidation} + E^{\circ}_{reduction}$$

$$\Delta E^{\circ} = 0.340 + 0.126 = 0.466\,V$$

The overall reaction is

$$Cu^{2+} + Pb \leftrightarrows cu + Pb^{2+}$$

$$\Delta E = \Delta E^{\circ} - \frac{RT}{nF}\ln(K)$$

$$\ln\frac{\left[Pb^{2+}\right]^{1}}{0.6M} = \frac{-(0.507v - 0.466v)(2)\left(\dfrac{96500C}{mole}\right)}{\left(8.134\dfrac{J}{mole\,K}\right)(298\,K)}$$

$$\left[Pb^{2+}\right] = 0.6\,M\exp(-3.1937) = 0.0246M$$

Chapter 4
Pourbaix Diagram

A Pourbaix diagram, also known as a potential/pH diagram, illustrates the electrochemical behavior of a metal in water at room temperature. It maps the relationship between electrode potential and pH and indicates the stability regions for the metal, including corrosion, immunity, and passivation. For instance, the behavior of nickel in water is shown in Fig. 4.1.

The construction of a Pourbaix diagram is based on predicted chemical reactions between the metal and the aqueous environment. These reactions define the equilibrium boundaries between different stability phases. When nickel (Ni) is immersed in water, several possible reactions may occur due to its interaction with H_2O, depending on the pH and potential conditions.

1. $Ni^{++} + 2e \rightarrow Ni$ depends only on the potential (due to the presence of e).
2. $NiO + 2H^+ + 2e \rightarrow Ni + H_2O$ depends on the pH & potential (due to the presence of e&H^+).
3. $Ni + 2H_2O \rightarrow Ni(OH)_2 + 2H^+$ depends only on the pH (due to the presence of H^+).
4. $NiO + H_2O \rightarrow Ni(OH)_2$ independent on both pH & potential (due to the absence of e&H^+).
5. $H^+ + e \rightarrow (1/2)H_2$ depends on the pH &potential (due to the presence of e&H^+).
6. $O_2 + 4H^+ + 4e \rightarrow 2H_2O$ depends on the pH &potential (due to the presence of e&H^+).
7. $O_2 + 2H_2O + 4e \rightarrow 4OH^-$ depends on the pH & potential(due to the presence of e&OH^-).
8. $2H_2O + 2e \rightarrow H_2 + 2OH^-$ depends on the pH & potential(due to the presence of e&OH^-).

The first step in constructing a Pourbaix diagram is to gather the standard electrode potentials for all relevant chemical species involved in the system. These standard potentials serve as the foundation for predicting the stability regions and plotting the equilibrium lines between corrosion, passivation, and immunity zones.

M. M. Ghatus, *Fundamentals of Corrosion Science and Engineering*,
https://doi.org/10.1007/978-3-032-13138-6_4

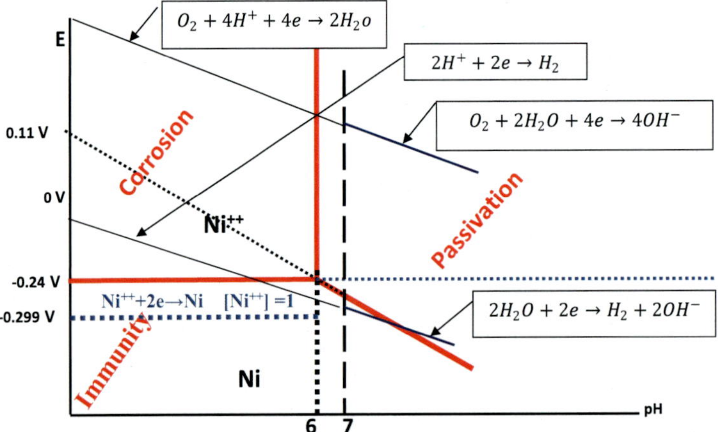

Fig. 4.1 Pourbaix diagram

$$\mu^{\circ}_{Ni^{++}} = -46398\,\text{J / mole}$$

$$\mu^{\circ}_{Ni} = 0\,\text{J / mole}$$

$$\mu^{\circ}_{NiO} = -215729.8\,\text{J / mole}$$

$$\mu^{\circ}_{Ni(OH)_2} = -452694\,\text{J / mole}$$

$$\mu^{\circ}_{H^+} = 0\,\text{J / mole}$$

$$\mu^{\circ}_{H_2} = 0\,\text{J / mole}$$

$$\mu^{\circ}_{O_2} = 0\,\text{J / mole}$$

$$\mu^{\circ}_{H_2O} = -236964.2\,\text{J / mole}$$

$$\mu^{\circ}_{OH^-} = -157147.1\,\text{J / mole}$$

Constructing a Pourbaix diagram (potential vs. pH) requires calculating the Gibbs free energy (ΔG°) for each relevant reaction. These values are then used to determine the standard electrode potentials (E°) for all reactions that depend on electrochemical potential..

For the reaction number (1)

$$Ni^{++} + 2e \rightarrow Ni$$

$$E_{\frac{Ni^{++}}{Ni}} = E^{\circ}_{\frac{Ni^{++}}{Ni}} + \frac{RT}{2F}\ln\frac{\left[aNi^{++}\right]}{\left[aNi\right]=1}$$

$$E_{\frac{Ni^{++}}{Ni}} = E^{\circ}_{\frac{Ni^{++}}{Ni}} + \frac{RT*2.303}{2F}\log\left[Ni^{++}\right]$$

$$E_{\frac{Ni^{++}}{Ni}} = E^{\circ}_{\frac{Ni^{++}}{Ni}} + \frac{0.05912}{2}\log\left[Ni^{++}\right]$$

$$E_{\frac{Ni^{++}}{Ni}} = E^{\circ}_{\frac{Ni^{++}}{Ni}} + 0.02956\log\left[Ni^{++}\right]$$

$$\left[Ni^{++}\right]=1$$

$$E_{\frac{Ni^{++}}{Ni}} = E^{\circ}_{\frac{Ni^{++}}{Ni}} \quad (***)$$

$$\Delta G^{\circ} = \mu^{\circ}_{product} - \mu^{\circ}_{reactent}$$

$$\Delta G^{\circ} = \mu^{\circ}_{Ni} - \mu^{\circ}_{Ni^{++}}$$

$$\Delta G^{\circ} = 0 - \left(-46398\frac{J}{mole}\right) = 46398\frac{J}{mole}$$

$$\Delta G^{\circ} = -nFE^{\circ}$$

$$E^{\circ} = -\frac{\Delta G^{\circ}}{nF} = -\frac{46398}{2*96500} = -0.24\,V$$

Substitute in equation ***

$$E_{\frac{Ni^{++}}{Ni}} = -0.24\,V$$

Since Reaction (1) depends solely on the electrode potential and is independent of pH, its representation in the Pourbaix diagram appears as a horizontal line (i.e., parallel to the pH axis). This line reflects the equilibrium potential for that reaction.

Although the line remains horizontal, the position of the line (vertically) will shift upward or downward depending on the concentration of nickel ions. An increase in Ni^{2+} concentration raises the equilibrium potential, while a decrease lowers it.

$$\text{If}\left[Ni^{++}\right]=10^{-3}$$

The equations (***) become as follows:

$$E_{\frac{Ni^{++}}{Ni}} = E^{\circ}_{\frac{Ni^{++}}{Ni}} - 0.02956 * 2$$

$$E_{\frac{Ni^{++}}{Ni}} = -0.24 - 0.059$$

$$E_{\frac{Ni^{++}}{Ni}} = -0.299v$$

The horizontal line in the Pourbaix diagram will shift downward as the concentration of Ni^{2+} ions decreases. In other words, the stability region where nickel exists as a metal becomes narrower. When the concentration falls below 10^{-6} M, the metal's presence becomes thermodynamically insignificant and may be considered unmeasurable in practice. For the reaction number (2)

$$\text{NiO} + 2\text{H}^{+} + 2\text{e} \rightarrow \text{Ni} + \text{H}_2\text{O}$$

$$E_{\frac{NiO}{H^+}} = E^{\circ}_{\frac{NiO}{H^+}} + \frac{RT}{2F}\ln\frac{[NiO]\left[H^+\right]^2}{[Ni][H_2O]}$$

$$E_{\frac{NiO}{H^+}} = E^{\circ}_{\frac{NiO}{H^+}} + \frac{RT}{2F}\ln\frac{1*\left[H^+\right]^2}{1}$$

$$E_{\frac{NiO}{H^+}} = E^{\circ}_{\frac{NiO}{H^+}} + \frac{0.059}{2}\log\left[H^+\right]^2$$

$$E_{\frac{NiO}{H^+}} = E^{\circ}_{\frac{NiO}{H^+}} - 0.059\left(-\log\left[H^+\right]\right)$$

$$E_{\frac{NiO}{H^+}} = E^{\circ}_{\frac{NiO}{H^+}} - 0.059\,\text{pH}$$

$$\Delta G^{\circ} = \mu^{\circ}_{product} - \mu^{\circ}_{reactent}$$

$$\left(0 - 236964.2\,\text{J/mole}\right) - \left(-215729.8\,\text{J/mole}\right) = -21234.4\,\text{J/mole}$$

$$E^{\circ} = -\frac{\Delta G^{\circ}}{nF} = -\frac{-21234.4}{2*96500} = 0.11\,\text{V}$$

$$E_{\frac{NiO}{H^+}} = 0.11 - 0.0295\,\text{pH}$$

At the pH of zero, the potential is equal to 0.11 V.

The potential is equal to −0.24 V; the pH is equal to 6.

Since nickel is immersed in water, the presence of species such as dissolved oxygen (O_2), water (H_2O), hydrogen gas (H_2), and hydroxide ions (OH^-) must be considered, as they can all participate in cathodic reactions. These species influence the electrochemical environment and help define the boundaries of corrosion, passivation, and immunity in the Pourbaix diagram.When the pH is greater than or equal to 7

$$2H_2O + 2e \rightarrow H_2 + 2OH^-$$

$$\Delta G^\circ = \mu^\circ_{product} - \mu^\circ_{reactent}$$

$$2(-157147.1\,\text{J/mole}) - 2(-236964.2\,\text{J/mole}) = 159634.2(\,\text{J/mole})$$

$$E = E^\circ + \frac{0.059}{2}\log\frac{[H_2O]^2}{[PH_2][OH^-]^2}$$

$$E° = -\frac{\Delta G^\circ}{nF} = -\frac{159634.2}{2 * 96500} = -0.827\,V$$

$$pOH = -\log[OH^-]$$

$$E = E^\circ + \frac{0.059}{2}\log\frac{[H_2O]^2}{[PH_2][OH^-]^2}$$

$$E = E^\circ + 0.059\log\frac{H_2O}{[PH_2][OH^-]^2}$$

$$pH + pOH = 14$$

$$E = E^\circ + 0.059(14 - pH)$$

$$E = -0.827 + 0.059 * 14 - 0.059\,pH$$

$$E = -0.059\,pH$$

$$O_2 + H_2O + 4e \rightarrow 4OH^-$$

$$E = 0.401 + 0.059 * 14 - 0.059pH$$

$$E = 1.22714 - 0.059pH$$

When the pH is less than 7

$$H^{+} + 2e \rightarrow H_2$$

$$E = -0.059pH$$

$$O_2 + 4H^{+} + 4e \rightarrow 2H_2o$$

$$E = 1.22714 - 0.059pH$$

Chapter 5
Exchange Current Density, Polarization, Activation Polarization, and Tafel Equation

5.1 Polarization

Polarization refers to the deviation of an electrode's potential from its equilibrium potential (defined by the exchange current density) due to net current flow. This shift occurs when an electrochemical reaction is driven away from equilibrium, causing a buildup or depletion of charge at the electrode surface.

5.2 Relationship Between Current Density and Free Energy

$$\vec{r}_1 = n_1 \gamma_1 f_1 \exp\left(-\frac{\Delta G}{RT}\right)$$

$$\overleftarrow{r}_2 = n_2 \gamma_2 f_2 \exp\left(-\frac{\Delta G}{RT}\right)$$

$\vec{r}_1$ is the flux of metal (mole/cm^2/sec) going to form metal ions

$\overleftarrow{r}_2$ is the flux of metal ion (mole/cm^2/sec) going to form metal atoms
n = moles/area,
γ = constant equal to 10^{13}/sec,
f = fraction of moles.

At the equilibrium condition, Fig. 5.1 at $\vec{r}_1$ (i_a) = $\overleftarrow{r}_2$ (i_c) number of moles/area/time.

$$n_1 \gamma_1 f_1 = n_2 \gamma_2 f_2$$

M. M. Ghatus, *Fundamentals of Corrosion Science and Engineering*,
https://doi.org/10.1007/978-3-032-13138-6_5

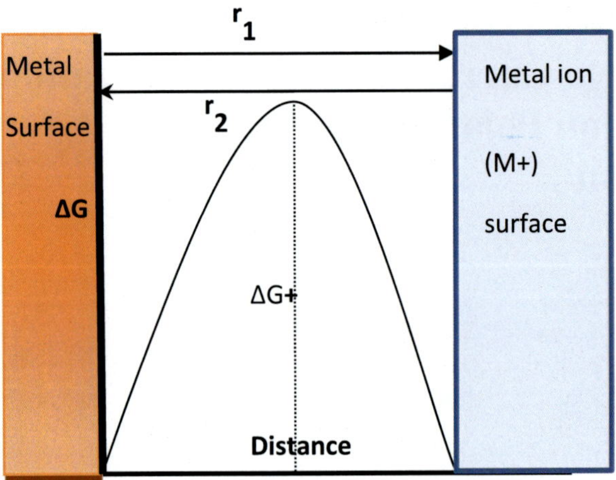

Fig. 5.1 Equilibrium condition

$$\frac{dn^{-}}{Adt} = \frac{1}{AnF}\left(\frac{dQ}{dt}\right)$$

where:

$\frac{dn^{-}}{Adt}$ is the number of moles transferred from the metal surface to the ion surface

$\frac{dQ}{dt}$ is the charge transfer rate.

$\frac{1}{A}\left(\frac{dQ}{dt}\right) = i_o$ represents the current density under equilibrium conditions at the metal surface. It is known as the exchange current density, and under this condition, no net corrosion occurs because the rates of the anodic and cathodic reactions are equal.

In other words, at equilibrium, the rate at which metal ions are released from the metal surface is equal to the rate at which metal ions are reduced and deposited back as metal atoms. This dynamic balance results in no net current flow, allowing the system to remain at its equilibrium potential.

$$\frac{dn^{-}}{Adt} = \frac{i_0}{nF}$$

$$\vec{r_1} = \overleftarrow{r_2} = \frac{i_0}{nF}$$

$$\frac{i_0}{nF} = n_1 \gamma_1 f_1 \exp\left(-\frac{\Delta G}{RT}\right)$$

$$i_o = nFn_1\gamma_1 f_1 \exp\left(-\frac{\Delta G}{RT}\right)$$

$$nFn_1\gamma_1 f_1 = A^-$$

$$i_o = A^- \exp\left(-\frac{\Delta G^\circ}{RT}\right)$$

$$i_a = A^- \exp\left(-\frac{\Delta G_a}{RT}\right)$$

$$i_c = A^- \exp\left(-\frac{\Delta G_c}{RT}\right)$$

When an external potential is applied, a positive potential leads to anodic polarization (G_a). In this case, the Gibbs free energy change (ΔG) becomes negative, indicating a spontaneous anodic reaction, as shown in Fig. 5.2.

Conversely, in the case of cathodic polarization (Gc), the applied potential is negative, and the direction of the free energy change is reversed.

$$G_a = -nF(E) = -ve$$

$$G_c = -nF(-E) = +ve$$

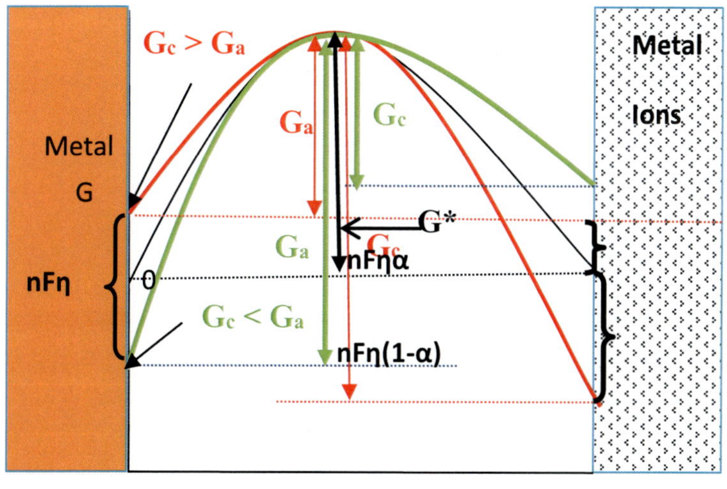

Fig. 5.2 The anodic and cathodic free energy

If $G_a > Gc$, then the cathodic current (i_c) is greater than the anodic current (i_a).

If $G_a < Gc$, then the anodic current (i_a) exceeds the cathodic current (i_c).

In both cases, the difference between the anodic and cathodic currents represents the applied ($i_{applied}$).

Overvoltage (Ƞ)

Overvoltage refers to the condition in which the equilibrium potential is disturbed due to the application of an external potential. It represents the extra voltage required to drive an electrochemical reaction at a rate different from equilibrium, resulting in either anodic or cathodic polarization.(±Ƞ).

- At the equilibrium condition:

  $M^{n+} + ne \rightleftarrows M$ the potential in this condition is equal to E°.

- After disturbing the equilibrium condition.

  M- ne → M^{n+} the potential in this condition is equal to E^-.

Case 1

(anodic polarization Ƞa or anodic overvoltage)

$$\vec{r_1} < r_{2\leftarrow}$$

$$|i_a| > |i_c| \rightarrow E^- - E^0_{\frac{m^{n+}}{m}} = +quantity$$

$$i_a - i_c = positive\ quantity\ \eta_a$$

The free energy will be negative (−ve).

Case 2

(cathodic polarization Ƞc or cathodic overvoltage)

$$\vec{r_1} > \overleftarrow{r_2}$$

$$|i_a| < |i_c| \rightarrow E^- - E^0_{\frac{m^{n+}}{m}} = -quantity$$

$$i_a - i_c = Negative\ quantity\ \eta_c$$

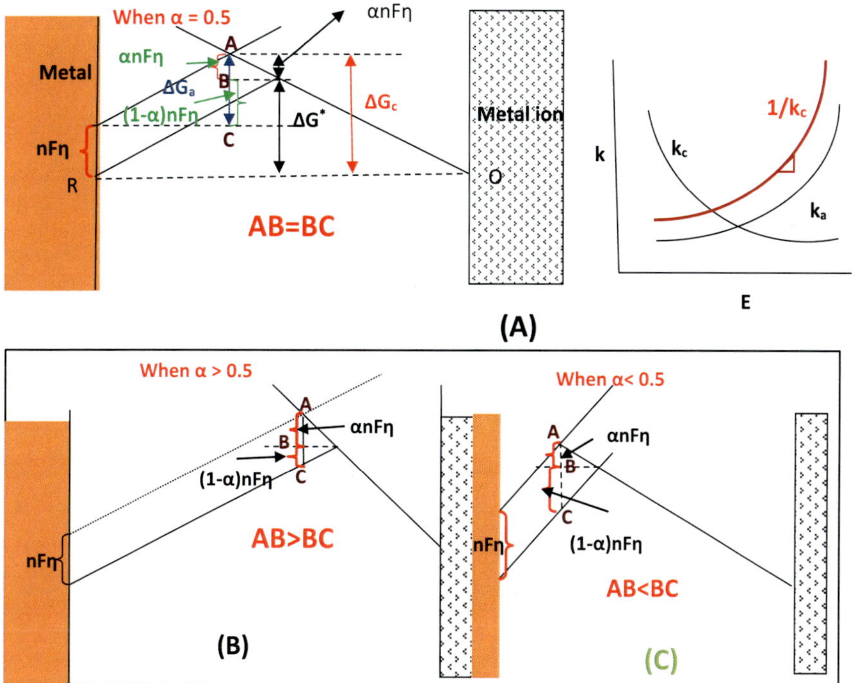

Fig. 5.3 Correlation between current density and overvoltage (different α ratio)

5.3 Correlation Between Current Density and Overvoltage

After polarization disturbs the equilibrium condition, the relationship between free energy and reaction coordinate (distance) changes. This shift in the activation energy barrier is illustrated in Figure 5.3a, where the energy profile reflects the increased or decreased difficulty of the electrochemical reaction depending on the direction of polarization.

When α (symmetry factor) = 0.5.

$$\mathrm{M^{n+} + ne \rightleftarrows M} \qquad \text{the potential in this condition } E^{o}$$

- After disturbing the equilibrium condition.

$$\mathrm{M \rightarrow M^{n+} + ne}$$

The potential in this condition E^{-}.

$$\text{If} \quad i_a > i_c \rightarrow \Delta G_c > \Delta G_a$$

$$r_a = n_1 \gamma_1 f_1 \exp\left(-\frac{\Delta G}{RT}\right)$$

$$if\ A'' = n_1\gamma_1 f_1\ r_a = A'' \exp\left(-\frac{\Delta G}{RT}\right)$$

$$i_a - |i_c| = i_{applied}$$

$$\text{Oxidant} + \text{ne} \leftrightarrows \text{Reductant}$$

$$\text{O} + \text{ne} \rightarrow \text{R}$$

K_c and K_a are backward and forward reactions, respectively
$r_c = k_c C_o$where C_o is the concentration of oxidant.
$r_a = k_a C_R$where C_R is the concentration of reductant.

$$E = E^\circ + \frac{RT}{nF}\ln\frac{C_o}{C_R}$$

$$E - E^\circ = \frac{RT}{nF}\ln\frac{C_o}{C_R}$$

When $r_a = r_c$

$$\text{k}_c\text{C}_o = \text{k}_a\text{C}_R$$

$$\ln k_c + \ln c_o = \ln k_a + \ln c_R$$

$$\ln k_a - \ln k_c = \ln\frac{c_o}{c_R}$$

$$\ln k_a + \ln\frac{1}{k_c} = \ln\frac{c_o}{c_R}$$

$$\ln k_a + \ln\frac{1}{k_c} = \frac{nF\left(E - E^\circ\right)}{RT}$$

By differentiating both sides with respect to E

$$\frac{d}{dE}\ln k_a + \frac{d}{dE}\ln\frac{1}{k_c} = \frac{d}{dE}\left(\frac{nF\left(E - E^\circ\right)}{RT}\right)$$

If n = 1

$$\frac{RT}{F}\frac{d}{dE}\ln k_a + \frac{RT}{F}\frac{d}{dE}\ln\frac{1}{k_c} = 1$$

$$\frac{RT}{F}\frac{d}{dE}\ln\frac{1}{k_c} \to \text{Redusive symmetry factor}(\alpha)$$

$$\frac{RT}{F}\frac{d}{dE}\ln k_a \to \text{Oxidative symmetry factor}(1-\alpha)$$

The total polarization is the summation of anodic and cathodic polarization

$$\eta(\text{total polariztion}) = \eta_a(\text{anodic polarization}) + \eta_c(\text{cathodic polarization})$$

If α (symmetry factor) is the fraction of total polarization or the amount of deviation referred to equilibrium status, so η_c **= α,** then $\eta_a = (1 - \alpha)$.

$$\Delta G_c = \Delta G^* + \alpha nF\eta$$

The change in activation energy for the anodic side.

$$\Delta G_a = \Delta G^* - nF\eta + \alpha nF\eta$$

$$\Delta G_a = \Delta G^* - nF\eta(1-\alpha)$$

The same principles apply when the symmetry factor (α) is either less than or greater than 0.5. The shape and height of the activation energy barrier are affected accordingly. These variations are illustrated in Figure 5.3 b and c, which show the energy profiles for cases, where $\alpha < 0.5$ and $\alpha > 0.5$, respectively.

5.4 Tafel's Equation

Tafel's equation is a fundamental expression in electrochemical kinetics that relates the rate of an electrochemical reaction to the overpotential applied to the system. It provides insight into how deviations from equilibrium affect current density.

Figure 5.4 illustrates this relationship, showing how the electrode potential varies linearly with the logarithm of the current density (log i) over a certain range of polarization.

$$A'' = n\gamma F$$

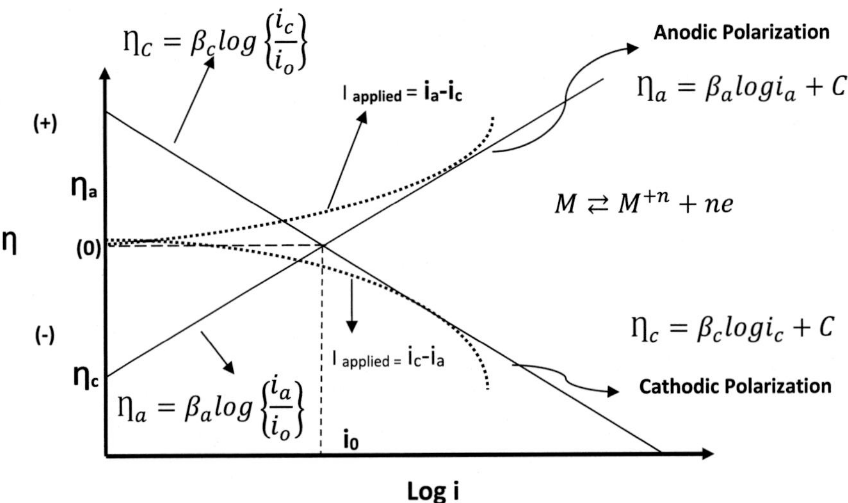

Fig. 5.4 Tafel's extrapolation

$$\frac{i_a}{nF} = A'' \exp\left(-\frac{\Delta G_a}{RT}\right)$$

$$i_0 = nFA'' \exp\left(\frac{-\Delta G^*}{RT}\right)$$

$$i_a = nFA'' \exp\left(-\frac{\Delta G^* - \alpha nF\eta}{RT}\right)$$

See Fig. 5.2.

$$i_a = nFA'' \exp\left(\frac{-\Delta G^*}{RT}\right)\exp\left(\frac{\alpha nF\eta}{RT}\right)$$

$$i_a = i_o \exp\left(\frac{\alpha nF\eta}{RT}\right)$$

$$i_c = nFA'' \exp\left(-\frac{\left(\Delta G^* + (1-\alpha)nF\eta\right)}{RT}\right)$$

See Fig. 5.2.

$$i_c = nFA'' \exp\left(\frac{-\Delta G^*}{RT}\right) \exp\left(-\frac{(1-\alpha)nF\eta}{RT}\right)$$

$$i_c = i_0 \exp\left(-\frac{(1-\alpha)nF\eta}{RT}\right)$$

$$i_{applied} = i_a - i_c$$

$$i_{applied} = i_o\left(\exp\left(\frac{\alpha nF\eta}{RT}\right) - \exp\left(-\frac{(1-\alpha)nF\eta}{RT}\right)\right) \rightarrow Butler-Volmer\ equation$$

$$\eta = E - E^{\circ}$$

When the polarization (η) value is very high, either in the cathodic or in the anodic direction.

If η is large positive $i_a = i_{appl}$

$$i_a = i_o \exp\left(\frac{\alpha nF\eta_a}{RT}\right)$$

If η is large negative $i_c = i_{appl}$

$$i_c = i_0 \exp\left(-\frac{(1-\alpha)nF\eta_c}{RT}\right)$$

When the polarization (η) or overvoltage is low

$$i_{applied} = i_o\left(\exp\left(\frac{\alpha nF\eta}{RT}\right) - \exp\left(-\frac{(1-\alpha)nF\eta}{RT}\right)\right)$$

$$\frac{i_a}{i_o} = \exp\left(\frac{\alpha nF\eta_a}{RT}\right)$$

$$\eta_a = \left\{\frac{RT}{+\propto nF}\right\} In\left\{\frac{i_a}{i_o}\right\}$$

$$\eta_a = \left\{\frac{RT\ 2.303}{+\propto nF}\right\} \log\left\{\frac{i_a}{i_o}\right\}$$

$$\beta_a = \left\{ \frac{RT\ 2.303}{+\propto nF} \right\} = \text{Anodic Tafel slope}\left(\text{positive slope}\right)$$

$$\left(\text{Slope between overvoltage and} \log i_a\right)$$

$$\eta_a = \beta_a \log \left\{ \frac{i_a}{i_o} \right\} = \beta_a \log i_a - \beta_a \log i_o$$

$$\eta_a = \beta_a \log i_a + C \ \textit{Anodic Tafel's equation}$$

where C = constant in Tafel's Equation

$$i_c = i_0 \exp\left(-\frac{(1-\alpha) nF\eta_c}{RT} \right)$$

$$\frac{i_a}{i_o} = \exp\left(-\frac{(1-\alpha) nF\eta_c}{RT} \right)$$

$$\eta_c = \left\{ -\frac{RT\ 2.303}{(1-\propto) nF} \right\} \log \left\{ \frac{i_c}{i_o} \right\}$$

$$\beta_c = \left\{ \frac{RT\ 2.303}{(1-\propto) nF} \right\} \textit{Cathodic Tafel slope}\left(\textit{negative slope}\right),$$

$$\left(\textit{Slope between overvoltage and} \log ic\right)$$

$$\eta_c = \beta_c \log \left\{ \frac{i_c}{i_o} \right\} = \beta_c \log i_c - \beta_c \log i_o$$

$$\eta_c = \beta_c \log i_c + C' \ \text{Cathodic Tafel's equation}$$

where C′ = constant in Tafel's Equation.

$$i_a = -i_c = i_o \eta = 0 \left(\text{No current flow}\right)$$

$$i_{net} = -i_{measured} = |i_a| - |i_c| = 0$$

Chapter 6
Mixed-Potential Theory

The mixed-potential theory accounts for both anodic and cathodic polarizations in electrochemical systems. It incorporates the influence of species diffusion in the electrolyte, which directly affects the current flowing through the system.

1- Corrosion involves:

   a. One or more oxidation reactions (anodic)
   b. One or more reduction reactions (cathodic)

2- The rate of electron generation (due to oxidation) is equal to the rate of electron consumption (due to reduction). In other words, both processes occur on the same metallic surface, maintaining electrochemical balance.

For example, when zinc is immersed in deaerated hydrochloric acid (HCl), zinc undergoes oxidation while hydrogen ions are reduced at the same surface, as illustrated in Fig. 6.1.

Observations
As shown in Fig. 6.2, when zinc is immersed in a zinc sulfate solution with activity equal to one, its electrode potential is −0.75 V. Simultaneously, when platinum is immersed in a 1 N hydrochloric acid (HCl) solution, it functions as a hydrogen electrode, with a standard potential of 0.00 V.

If the two half-cells are electrically connected to form a galvanic cell, the overall cell potential is calculated as:

$$E\text{cell} = E\text{cathode} - E\text{anode} = 0.00 - (-0.75) = +\ 0.75\,\text{VE}$$

This positive value indicates a spontaneous electrochemical reaction, with electrons flowing from zinc (anode) to platinum (cathode).

Figure 6.3 illustrates the relationship between current and electrode potential for zinc immersed in hydrochloric acid (HCl). In this system, both the anodic (zinc oxidation) and cathodic (hydrogen ion reduction) reactions occur on the same zinc surface.

M. M. Ghatus, *Fundamentals of Corrosion Science and Engineering*,
https://doi.org/10.1007/978-3-032-13138-6_6

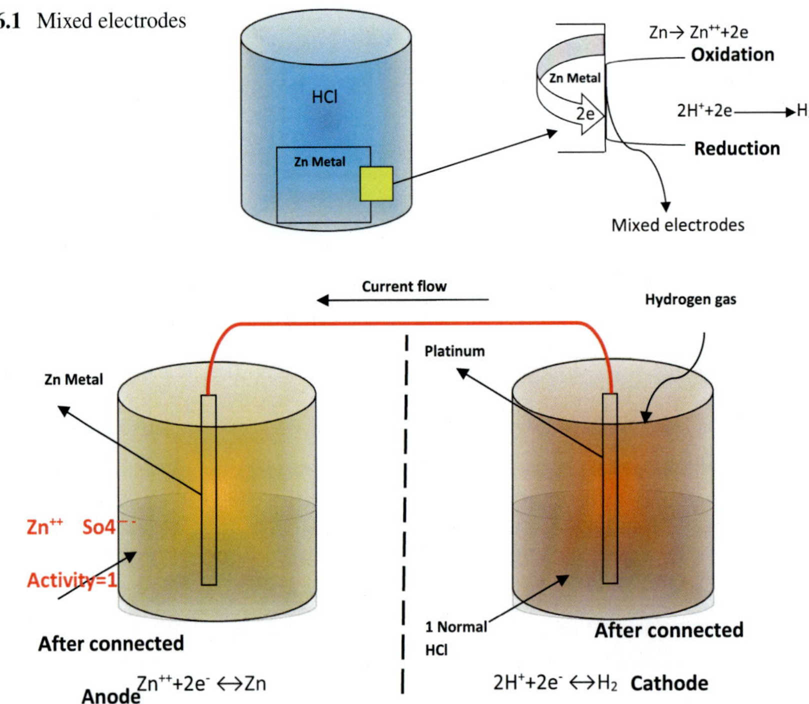

Fig. 6.1 Mixed electrodes

Fig. 6.2 Mixed-potential theory

As a result, the zinc gradually corrodes, while hydrogen gas evolves into visible bubbles. Both zinc and hydrogen reactions are reversible, with their respective exchange current densities being:

- i_0 (Zn) = 10^{-7} A/cm^2
- i_0 (H_2) = 10^{-11} A/cm^2

The significant difference in exchange current densities highlights the sluggish kinetics of hydrogen evolution on zinc, which contributes to localized corrosion behavior.

If iron is immersed in the same hydrochloric acid solution previously used for zinc, the corrosion rate of iron will be significantly higher than that of zinc under identical conditions:

$$i_{\text{corrosion,Fe}} \gg i_{\text{corrosion,Zn}}$$

This is because the exchange current density of hydrogen evolution (i_0) on iron is much greater than on zinc

$$i_0\left(\mathrm{H}_{\text{Zn surface}}\right) \ll i_0\left(\mathrm{H}_{\text{Fe surface}}\right)$$

In other words, hydrogen is more easily reduced on the iron surface, enhancing the cathodic reaction rate and thereby accelerating the overall corrosion of iron.

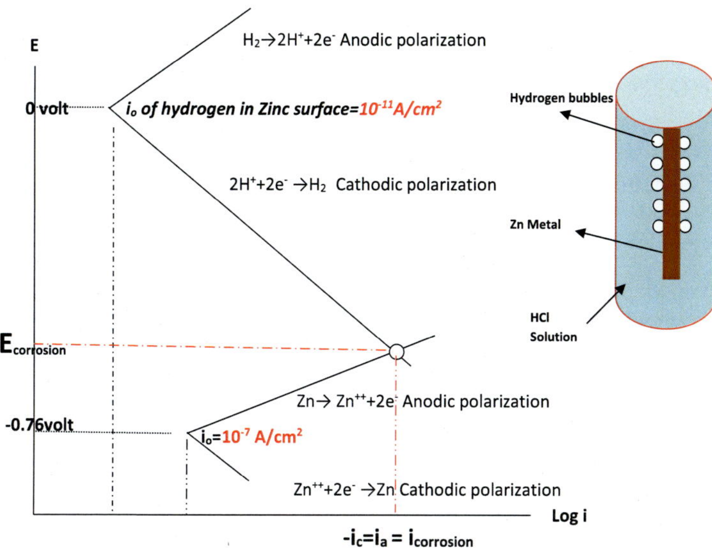

Fig. 6.3 The relation between the current and the potential for the zinc in HCl solution

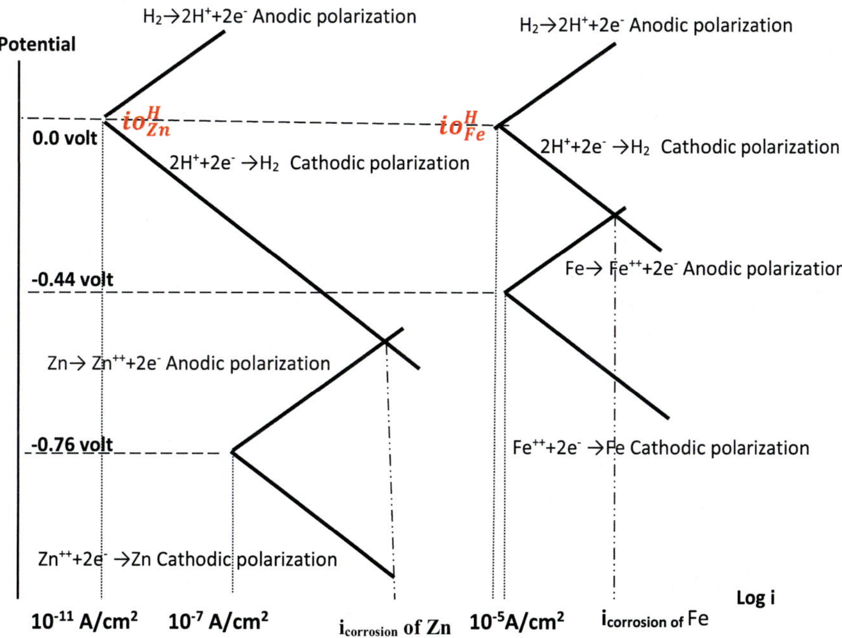

Fig. 6.4 Comparison between zinc and iron in the same solution

Figure 6.4 compares the corrosion behavior of zinc and iron in the same acidic environment, highlighting the effect of hydrogen kinetics on the corrosion process.

When iron and zinc are galvanically coupled, each with the same surface area and immersed in deaerated hydrochloric acid (HCl), the corrosion rate of zinc becomes significantly greater than that of iron. This indicates that zinc is actively corroding to protect the iron, acting as a sacrificial anode.

In Fig. 6.5, the corrosion current ($i_{\mathbf{corrosion}}$) and corrosion potential ($E_{\mathbf{corrosion}}$) are determined by superimposing:

- The anodic polarization curves of both zinc and iron, and
- The cathodic polarization curves for hydrogen evolution on both surfaces

The intersection point between the total anodic and total cathodic curves represents the combined corrosion current for the galvanic couple (iron + zinc) and defines the corrosion potential of the system.

To determine the individual corrosion current of either iron or zinc, draw a vertical line at the corrosion potential (Ecorrosion). The intersection of this line with the respective metal's anodic polarization curve gives the corrosion current for that metal.

To evaluate the galvanic behavior of zinc and iron, the following polarization curves must be considered:

- The anodic polarization curves of both zinc and iron
- The cathodic polarization curves for hydrogen evolution on both surfaces

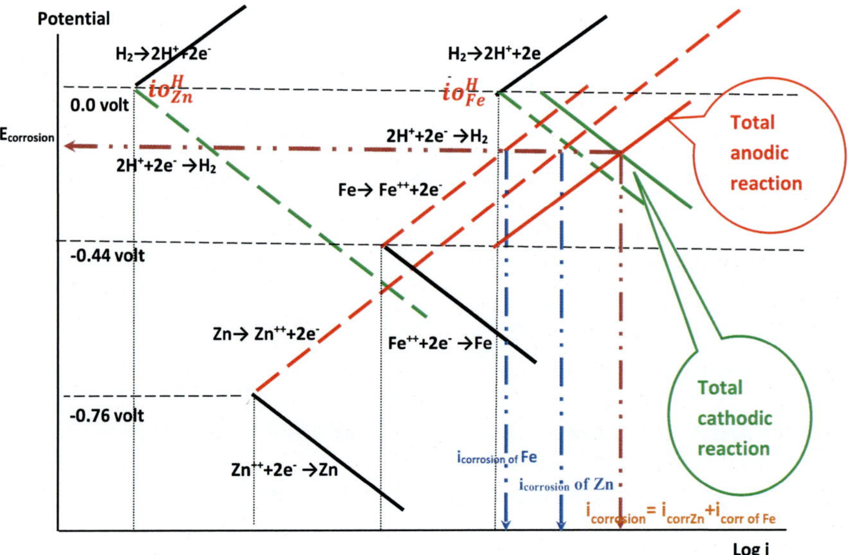

Fig. 6.5 Glavanically coupled Zn and Fe in deaerated HCl solution

By summing the individual anodic and cathodic currents, a set of total anodic and total cathodic curves is constructed. The intersection point of these two total curves defines the corrosion potential (Ecorrosion) and the overall corrosion current for the coupled system.

To determine the individual corrosion current for either iron or zinc, locate the point where the vertical line at Ecorrosion intersects the respective anodic polarization curve of that metal. This value represents the corrosion current ($i_{\mathbf{corrosion}}$) for either zinc or iron within the galvanic couple, as determined by their anodic response at the system's corrosion potential.

6.1 Explanation of Corrosion Events Based on Mixed-Potential Theory, the Effect of Impurities, and the Effect of Area Factor

6.1.1 Effect of Impurities

Zinc in Two Corrosive Environments.

A zinc rod is immersed in two different deaerated HCl solutions:

1. Case 1: Deaerated HCl solution (no impurities)
2. Case 2: Deaerated HCl solution containing ferric ions (Fe^{3+}) as impurities

The corrosion rate of zinc in Case 2 is significantly higher than in Case 1. Why?

In Fig. 6.6, the polarization diagram shows:

- Two cathodic reactions:
  - One due to hydrogen ion reduction ($H^+ + e^- \rightarrow H_2$).
  - One due to the reduction of ferric ions ($Fe^{3+} + e^- \rightarrow Fe^{2+}$).
- One anodic reaction: zinc oxidation ($Zn \rightarrow Zn^{2+} + 2e^-$).

The presence of ferric ions introduces an additional cathodic reaction, which increases the total cathodic current. According to the mixed-potential theory, the system must adjust so that total anodic and cathodic currents are equal. As a result, the corrosion potential shifts, and the anodic current (zinc dissolution) increases, leading to a higher corrosion rate.

The intersection between the hydrogen reduction reaction and the anodic polarization curve of zinc in deaerated HCl represents the corrosion current of zinc under these conditions. This intersection occurs at Point A in Fig. 6.6.

When ferric ions (Fe^{3+}) are introduced as impurities, a second cathodic reaction is added to the system. The combined cathodic curve, which includes both hydrogen and ferric ion reduction reactions, intersects the anodic zinc curve at a new point—Point B.

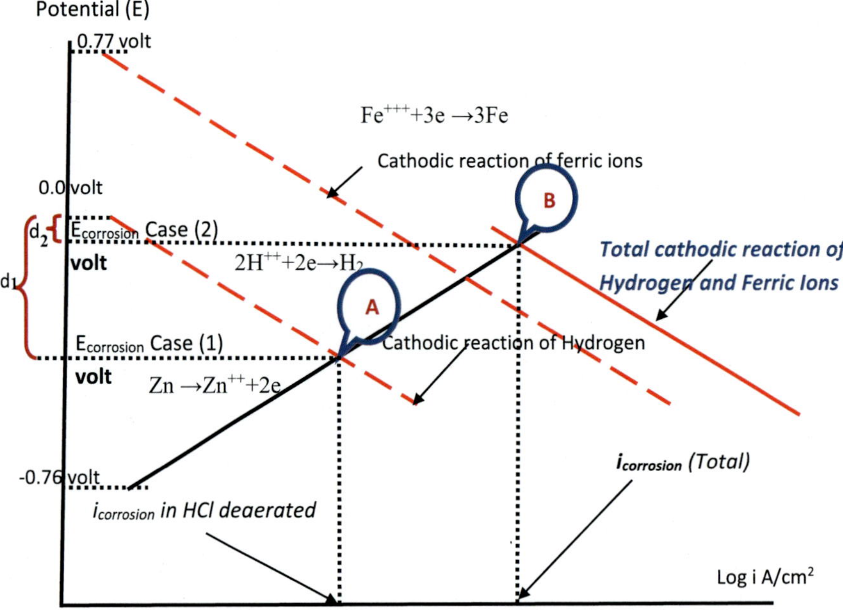

Fig. 6.6 The effect of impurities on the corrosion rate of Zn

This new intersection (Point B) represents the increased corrosion current of zinc in deaerated HCl containing ferric ion impurities, highlighting the accelerating effect of additional cathodic reactions on the overall corrosion process.

Therefore,

$$i_{catodic}^{Total} = i_{\frac{H^+}{H}} + i_{\frac{Fe^{+++}}{Fe^{++}}}$$

$$i_{anodic}^{Total} = i_{\frac{Zn^{++}}{Zn}}$$

At the intersection between the total cathodic reactions and the anodic reaction of zinc, point (B).

$$i_{catodic}^{Total} = i_{anodic}^{Zinc} = i_{corrosion}^{Case(2)}$$

Therefore, the **corrosion current in** Case 2 is greater than that in Case 1, indicating that the **corrosion rate of zinc** is **higher** when **ferric ion impurities** are present in the solution. In other words, more electrons are consumed in Case 2 due to the additional cathodic reaction, which drives a higher rate of zinc dissolution compared to the deaerated HCl solution alone.

In addition, ferric ions in deaerated HCl act as a depolarizer for the hydrogen evolution reaction. By introducing an additional cathodic pathway, they reduce the overpotential required for hydrogen reduction.

As shown in Fig. 6.6, the distance between the corrosion potential (Ecorr) in Case 2 and zero volts (d_2) is smaller than the corresponding distance d_1 in Case 1. This reduction in polarization confirms that ferric ions decrease the cathodic overpotential, thus acting as effective depolarizers and accelerating the overall corrosion process.

Figure 6.7 illustrates the effect of a noble metal on the corrosion rate of an active metal, a phenomenon known as galvanic corrosion.

In the electrochemical series, standard electrode potentials are measured relative to the standard hydrogen electrode (SHE), which is assigned a value of 0 volts. Metals with higher (more positive) standard potentials are considered nobler, while those with lower (more negative) potentials are more active.

For example:

- Gold (Au) has a standard potential of +1.50 V, making it highly noble.
- Platinum (Pt) has a standard potential of +0.242 V.
- Zinc (Zn) has a standard potential of −0.763 V, making it much more active than both platinum and gold.

As the standard potential increases relative to hydrogen, a metal's nobility increases, meaning it is less likely to corrode. Conversely, a more negative potential indicates higher activity and greater corrosion susceptibility, particularly when in electrical contact with a nobler metal.

If platinum or gold is electrically coupled with zinc in a deaerated HCl solution, the hydrogen reduction reaction occurs on both the noble metal surface (gold or platinum) and the zinc surface.

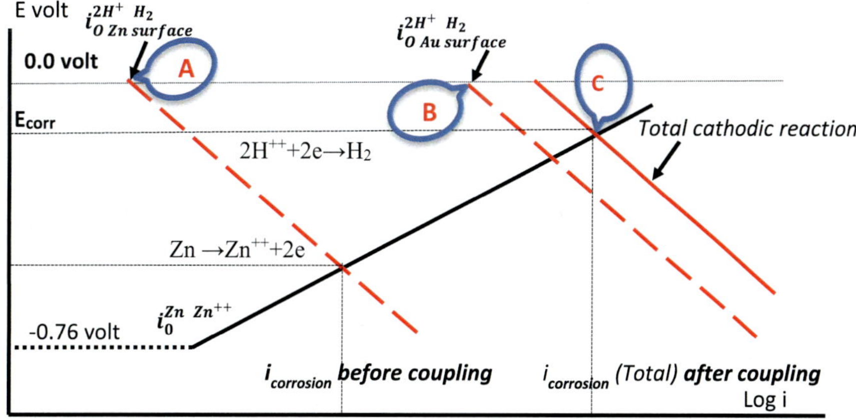

Fig. 6.7 The effect of noble metal on the corrosion rate of active metal

As shown in Fig. 6.7, when gold is connected to zinc, two cathodic hydrogen reduction reactions take place simultaneously—one on the zinc surface and one on the gold surface. Since gold is significantly more catalytically active for hydrogen evolution, it contributes substantially to the overall cathodic current.

When the currents from both cathodic sites are summed, the total cathodic current increases, and according to the mixed-potential theory, the anodic reaction (zinc dissolution) must increase accordingly. This results in a significant rise in the corrosion current of zinc when it is coupled with gold, thereby accelerating galvanic corrosion of zinc.

At point (A) is the cathodic current of hydrogen at the zinc surface $i_{C\,Zn\,surface}^{\frac{2H^+}{H_2}}$

At point (B) is the cathodic current of hydrogen at the gold surface $i_{C\,Au\,surface}^{\frac{2H^+}{H_2}}$

Point C on the polarization diagram represents the anodic current of zinc, which is equal to the total cathodic current resulting from hydrogen reduction occurring on both the gold and zinc surfaces. This balance satisfies the condition of electrochemical equilibrium within the galvanic couple.

$$i_a^{Zn \to Zn^{++}+2e} = -i_c\,Total \left(i_c\,Total = i_{C\,Zn\,surface}^{\frac{2H^+}{H_2}} + i_{C\,Au\,surface}^{\frac{2H^+}{H_2}} \right)$$

6.1.2 *Area Effect*

The surface area of connected metals plays a critical role in determining the corrosion rate in a galvanic couple. In such systems, the more active metal (anode) undergoes corrosion and sacrifices itself to protect the less active metal (cathode).

For example, if two copper plates are connected by a rivet made of iron and placed in seawater, the surface area of the copper is much larger than that of the iron rivet. According to the galvanic series in seawater, iron is more active than copper, meaning iron will act as the anode and corrode, while copper will be protected.

In this case, the corrosion process involves:

- One anodic reaction: Iron oxidation: $Fe \rightarrow Fe^{2+} + 2e$.
- One cathodic reaction (on the copper surface).
- Oxygen reduction in seawater (neutral/alkaline medium).

Because of the large cathodic area (copper) and small anodic area (iron), the current density on the iron becomes very high, accelerating its corrosion.

$$O_2 + 2H_2O + 4e^- \rightarrow 4OH^-$$

$$O_2 + 4H^+ + 4e \rightarrow 2H_2O \text{ Cathodic reaction}$$

$$Fe \rightarrow Fe^{++} + 2e \text{ Anodic reaction}$$

Based on this equation:

$$i_a = \frac{I}{A_a}$$

By considering the activation polarization, in other words, the cathodic and anodic reactions are both linear.

$$\eta_{activation} = \beta_{a/c} \log\left\{\frac{i_{a/c} * A_{a/c}}{i_o * A_{a/c}}\right\}$$

When the anodic area is significantly smaller than the cathodic area, the current required from the anode (i_a) becomes very high to balance the cathodic reaction. As a result, the current density on the anodic surface increases, which leads to a higher corrosion rate on the anode.

Therefore, when the anode area is smaller than the cathode area, the localized corrosion rate on the anode intensifies. Conversely, if the anode area is larger than the cathode area, the current is distributed over a wider surface, and the corrosion rate per unit area of the anode is reduced.

- Figure 6.8 illustrates the relationship between total current (I), measured in amperes (A), and electrode potential (E). In this figure, current density (A/cm^2) is intentionally omitted to emphasize the effect of electrode surface area on the corrosion current, and therefore on the corrosion rate.

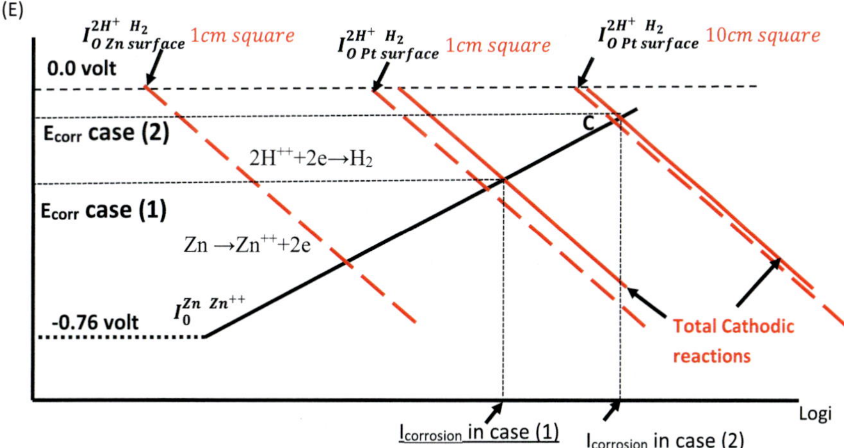

Fig. 6.8 The effect of surface area on the corrosion rate of zinc

The figure presents two cases for comparison:

- A galvanic couple with a small anodic area and a large cathodic area.
- A galvanic couple with a large anodic area and a small cathodic area.

These two cases highlight how area ratios directly influence the magnitude of total corrosion current and the severity of anodic metal dissolution.

- Case (1) the ratio between the area of zinc and platinum is 1:1.
- Case (2), the ratio between the area of zinc and platinum is 1:10, respectively.

Figure 6.8 shows two cathodic polarization curves for the hydrogen reduction reaction:

- One curve represents hydrogen evolution on a zinc surface with an area of 1 cm^2.
- The second curve represents the same reaction on a platinum surface, also with an area of 1 cm^2.

This is defined as Case (1), where the cathodic surface areas are equal for both zinc and platinum.

In Case (2), the surface area of the platinum is increased to 10 cm^2, while the zinc area remains at 1 cm^2. As a result, the cathodic polarization curve for platinum shifts to the right, indicating a higher total cathodic current at any given potential.

This increase in cathodic area causes the corrosion current (or corrosion rate) of the zinc (anode) to increase compared to Case (1). This behavior demonstrates a key principle.

The corrosion rate of the anode is directly proportional to the surface area of the cathode in a galvanic couple.

Chapter 7
Concentration Polarization

Concentration polarization refers to the portion of the total polarization in an electrolytic cell that arises from changes in the electrolyte concentration near the electrode/solution interface as a result of current flow. This occurs when the rate of electrochemical reaction at the surface exceeds the rate at which reactive species can be replenished by mass transport from the bulk solution.

Figure 7.1 illustrates the concentration gradient from the metal surface to the bulk electrolyte, showing the depletion of ions near the interface due to ongoing electrochemical reactions.

This behavior is governed by Fick's law of diffusion, which relates the current density to the concentration gradient in the electrolyte. It is an essential principle in quantifying how mass transport limitations affect electrochemical reaction rates.

$$i = -nFD\frac{\left(C_O - C\right)}{\delta}$$

When the current (i) reaches its maximum value, known as the limiting current, it indicates that the entire concentration of the reactive species at the metal surface has been consumed by the electrochemical reaction. At this point, the concentration at the electrode surface (C) becomes effectively zero, meaning no additional reduction can occur unless more species are transported from the bulk solution.

This condition reflects mass-transport-limited behavior, where the rate of the reaction is no longer controlled by electrode kinetics but by the diffusion rate of species to the surface.

$$i_{\max} = -nFD\frac{C_O}{\delta}$$

M. M. Ghatus, *Fundamentals of Corrosion Science and Engineering*,
https://doi.org/10.1007/978-3-032-13138-6_7

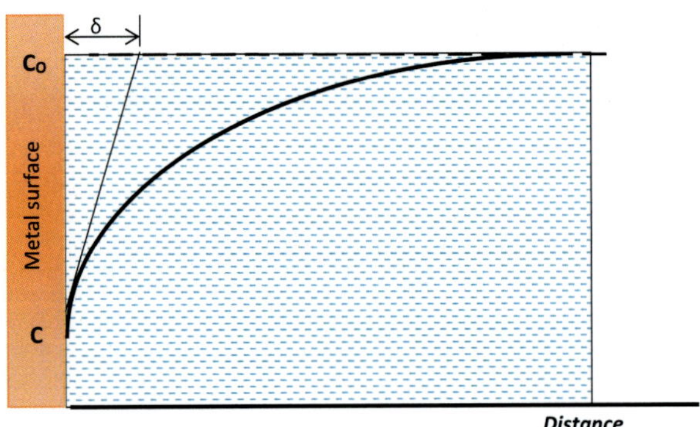

Fig. 7.1 The Concentration profile from the metal surface to the bulk solution

$$C_O = -\frac{i_{\max}}{nFD}\delta$$

where:

- **D** Diffusivity of the metallic species in the medium
- **F** Faraday's constant
- **n** number of electrons
- **δ** boundary layer
- **C** concentration Distance

The Nernst equation is used to evaluate the electrode potential at any position near the electrode/solution interface, based on the local concentration of ionic species. It provides a quantitative relationship between the electrode potential, the standard potential, and the activities or concentrations of the reactants and products involved in the electrochemical reaction.

$$E^{(1)}_{M^{n+}/M} = E^{\circ}_{M^{n+}/M} + \frac{RT}{nF} 2.303 \log C_0$$

$$E^{(2)}_{M^{n+}/M} = E^{\circ}_{M^{n+}/M} + \frac{RT}{nF} 2.303 \log C$$

When the concentration at the electrode surface (C) is less than the bulk concentration (C_o), that is, $C_o > C$, the result is a drop in potential. In other words, the local electrode potential

$$\mathrm{E}_2 < \mathrm{E}_1$$

$$\text{decreases}: \quad \eta_{\text{cath concentration}} = E^{(2)}_{M^{n+}/M} - E^{(1)}_{M^{n+}/M}$$

$$\eta_{\text{cath concentration}} = \frac{0.059}{n}\log\frac{C}{C_O}$$

$$\frac{C}{C_O} = 1 - 1 + \frac{C}{C_O} \qquad \frac{C}{C_O} = 1 - \frac{C_0 - C}{C_O}$$

$$C_O - C = -\frac{i\delta}{nFD}$$

$$\text{At}\,\text{C} = 0 \qquad C_O = -\frac{i_{\max}\delta}{nFD}$$

$$\frac{C}{C_O} = 1 - \frac{i}{i_{\max}}$$

$$\eta_{cath\ concentration} = \frac{0.059}{n}\log\left(1 - \frac{i}{i_{\max}}\right)$$

$$\eta_{conc} = E^{(2)}_{M^{n+}/M} - E^{(1)}_{M^{n+}/M} = \frac{0.059}{n}\log\left(1 - \frac{i}{i_{\max}}\right)$$

The relationship between log i and concentration polarization is:

$$\text{At}\,\text{i} = 0 \rightarrow \eta_{\text{concetration}} = 0$$
$$\text{At}\,\text{i} = \text{i}_{\text{maximum}} \rightarrow \eta_{\text{concetration}} = \infty$$

As shown in Fig. 7.2, when the current (i) is less than the maximum (limiting) current ($i_{\mathbf{max}}$), the effect of concentration polarization is minimal or nearly zero. Under these conditions, the system is operating in the activation-controlled region, where the electrode potential is primarily influenced by electrode kinetics, not mass transport.

Because the concentration near the electrode surface remains close to the bulk value, no significant depletion occurs, and the potential does not shift due to concentration effects. However, as the current increases toward $i_{\mathbf{max}}$, the potential begins to shift to the more negative side, indicating the onset of cathodic concentration polarization.

Because the concentration near the electrode surface remains close to the bulk value, no significant depletion occurs, and the potential remains stable, not affected by concentration polarization. However, as the current approaches the limiting value (i_{max}), the concentration at the surface begins to drop, and the potential shifts to the more negative side, marking the onset of cathodic concentration polarization.

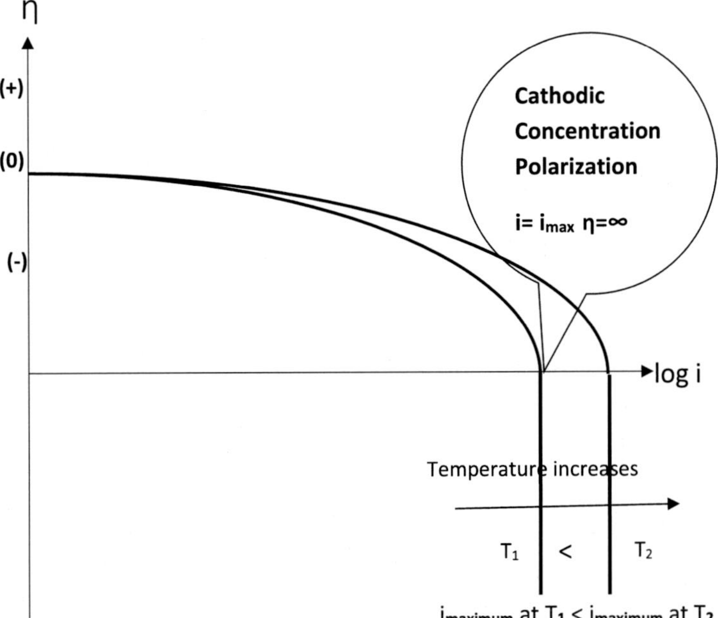

Fig. 7.2 The effect of temperature on concentration polarization

Furthermore, temperature has a direct effect on diffusion. As temperature increases, the diffusivity of ions also increases, which in turn leads to an increase in the limiting current (i_{max}).

For example, in aqueous solutions, the diffusivity of hydrogen ions (H^+) is greater than that of dissolved oxygen (O_2), because H^+ ions are significantly smaller and more mobile than oxygen molecules.

$$D_{O_2} \ll D_{H^+}$$

$$i_{max}^{O_2} \ll i_{max}^{H^+}$$

In a specific case, when the temperature increases from T_1 to T_2, the diffusivity of hydrogen ions (H^+) increases due to enhanced thermal motion. As a result, the limiting current (i_{max}) at T_2 becomes greater than at T_1, allowing a higher rate of electrochemical reaction.

When the applied current reaches i_{max} infinite polarization is theoretically observed. At this point, the concentration of reactive species at the electrode surface drops to zero, and the potential shifts drastically due to complete mass transport limitation.

In aqueous environments, two common cathodic reactions can occur:

- Dissolved oxygen reduction (in neutral or alkaline solution):

$$O_2 + 4H^+ + 4e \rightarrow 2H_2O \cong E^\circ = 1.023\,\text{Volt}$$

- Hydrogen ion reduction

$$H^{+} + e \rightarrow 0.5H_2 \cong E^{\circ} = 0.0\,\text{Volt}$$

Figure 7.3 illustrates the polarization behavior of the two major cathodic reactions in aqueous environments: oxygen reduction and hydrogen ion reduction. From this figure, the following key observations can be made:

- The activation polarization of the oxygen reduction reaction begins at approximately 1.023 volts.
- As current increases, the potential shifts negatively until it reaches the limiting current (i_{max}) for oxygen. Beyond this point, infinite concentration polarization occurs, and the potential rapidly drops toward 0 volts.
- At 0 volts, which is the standard potential (E°) for the hydrogen reduction reaction, a new activation polarization curve begins for hydrogen. This continues until it reaches the limiting current (i_{max}) for hydrogen.
- Beyond this point, hydrogen reduction also enters the infinite concentration polarization region.
- As previously discussed, the maximum current for hydrogen is greater than that for oxygen, due to the higher diffusivity and mobility of hydrogen ions compared to oxygen molecules in water.

Reasons Why Anodic Concentration Polarization Is Not Considered

1. Corrosion Product Formation and Passivation:
    - At the interface between the metal and the electrolyte, the formation of corrosion products leads to the passivation of the metal surface. This process decreases the rate of further corrosion by creating a protective oxide or other passive layer.

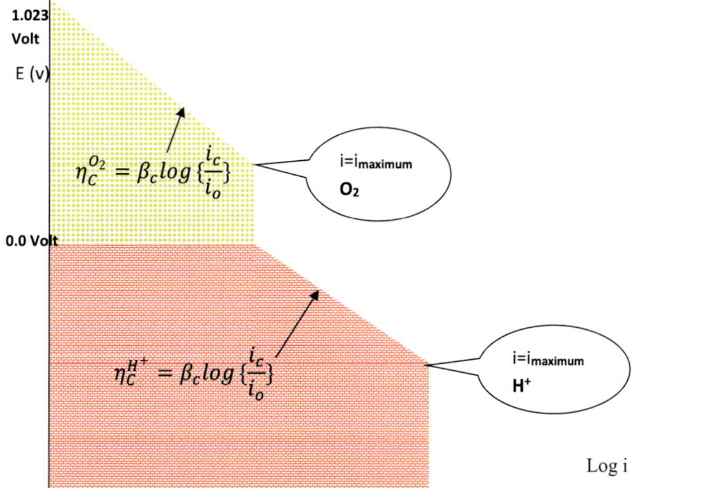

Fig. 7.3 Cathodic polarization of O_2 and H_2 and maximum current

2. Limiting Current in Reduction Processes:

   - The limiting current (i_{max} or $i_{limiting}$) represents the maximum current at which polarization becomes infinite. This is a phenomenon primarily significant during reduction reactions, such as the hydrogen evolution reaction or oxygen reduction in electrochemical processes.

3. Unlimited Supply of Metal Ions:

   - In electrochemical reactions, there is an unlimited supply of metal ions at the electrode interface. This assumption is key to understanding that, at higher currents, the limiting factor becomes the ability of the electrolyte to replenish the ions, particularly in the case of cathodic reactions like hydrogen evolution.

Conclusion: All Polarization Based on the Cathodic Reaction
From the above points, it can be concluded that all types of polarization (activation, concentration, and limiting current) are heavily influenced by the cathodic reactions in electrochemical systems. The characteristics of these reactions, especially the availability of ions and the diffusion limits, directly determine the behavior and magnitude of polarization

$$M^{+n} + ne \rightarrow M$$

$$\eta_{\text{activation}} = \beta_{a/c} \log\left\{\frac{i_{a/c}}{i_o}\right\}$$

$$\eta_{\text{cath concentration}} = -\frac{0.059}{n}\log\left(1 - \frac{i}{i_{\max}}\right)$$

$$\text{Resistance polarization } \eta_{\text{resistance}} = \text{IR} = \text{Volts}$$

$$\eta_{\text{Total}} = \eta_{\text{activation}} + \eta_{\text{cathodic concentration}} + \eta_{\text{Resistance}}$$

In 1938, Wagner and Traud introduced two fundamental hypotheses that form the basis of the mixed potential theory in corrosion science:

1- First Hypothesis—Mixed Potential Formation:

In any corrosion process involving multiple electrochemical reactions, a single, unique potential—called the mixed potential—is established at which the total anodic current equals the total cathodic current. This potential represents the point of electrochemical equilibrium on the corroding metal surface.

2- Second Hypothesis—Independence of Partial Reactions:

Each partial reaction (anodic or cathodic) proceeds independently on the electrode surface, as though it were the only reaction occurring, but its rate is governed by the overall mixed potential of the system.

These hypotheses provided the theoretical foundation for interpreting corrosion behavior using polarization curves, and they remain central to modern corrosion science.

Any electrochemical reaction can be divided into two or more partial oxidation and reduction reactions; for example, zinc in deaerated HCl

$$Zn \rightarrow Zn^{++} + 2e \qquad \text{Anodic reaction}$$

$$2H^{+} + 2e \rightarrow H_2 \qquad \text{Cathodic reaction}$$

The overall reaction for the above two reactions is

$$Zn + 2H^{+} \rightarrow Zn^{++} + H_2$$

If the HCl is aerated, another oxygen reduction reaction is added

$$O_2 + 4H^{+} + 4e \rightarrow 2H_2O$$

There can be no net accumulation of electric charge during an electrochemical reaction. In other words, according to the principle of charge conservation, the overall rate of electron consumption (i_c) in cathodic reactions must exactly match the overall rate of electron generation (i_a) in anodic reactions:

$$-i_c = i_a = i_{\text{corrosion}}$$

7.1 Concentration Polarization and Corrosion Current

The relationship between concentration polarization and the corrosion current (or limiting current) is described by the following equation

$$\eta_{\text{cath concentration}} = -\frac{0.059}{n}\log\left(1-\frac{i}{i_{\text{limiting}}}\right)$$

Figure 7.4 illustrates the relationship between concentration polarization and its effect on the corrosion rate.

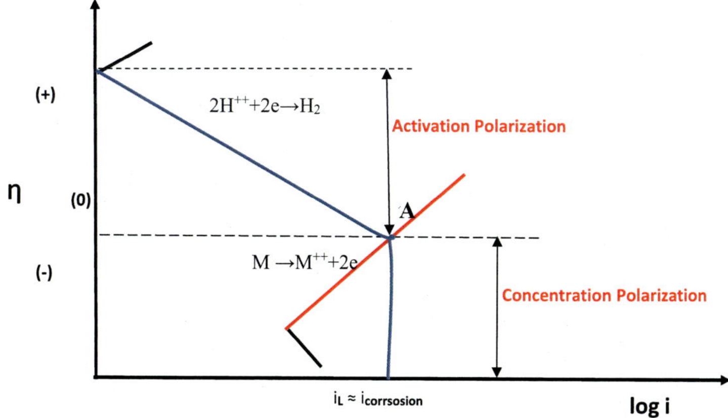

Fig. 7.4 The effects of concentration polarization on the value of the corrosion rate

For example, if the metal dissolution (anodic reaction) curve intersects with the cathodic concentration polarization curve, the point of intersection determines the system's corrosion potential (Ecorr) and corrosion current density (icorr).

In this case, the cathodic reaction limits the corrosion rate, specifically by the mass transport of the species being reduced (e.g., oxygen or hydrogen ions). The closer the system operates to the limiting current, the more significant the concentration polarization becomes, reducing the efficiency of cathodic processes and increasing the required anodic current to maintain electrochemical balance.

As a result, concentration polarization can restrict the cathodic reaction rate, thereby limiting the overall corrosion rate, even when the anodic reaction is kinetically favorable.

$$M - 2e \rightarrow M^{++}$$

- At point (A) $i_{\mathbf{corrosion}} = i_{\mathbf{limiting}}$ = concentration polarization infinite ($\eta_{\mathbf{infinite}}$).

7.2 Effect of Velocity

The effect of fluid velocity on the limiting current or corrosion current is governed by mass transport dynamics at the electrode surface.

As the fluid velocity increases, the thickness of the diffusion (boundary) layer (δ) at the cathodic surface decreases (as shown in Fig. 7.1).

A thinner boundary layer allows ions or reactive species (e.g., O_2 or H^+) to reach the electrode surface more rapidly, thus increasing the limiting current ($i_{\mathbf{lim}}$).

This behavior is described by the following equation derived from Fick's first law:

$$i_L = \frac{DnFC_0}{\delta}$$

Figure 7.5 illustrates the effect of fluid velocity on both the limiting current and activation polarization. This effect can be categorized based on the magnitude of velocity as follows:

Stagnant Flow

When the fluid velocity is zero, mass transport is minimal, the diffusion layer (δ) is thick, and the system is strongly influenced by concentration polarization. The limiting current is very low, and the corrosion process is significantly limited by species availability at the electrode surface.

Increasing Velocity (V_1, V_2, … V_n)

As the velocity gradually increases, the diffusion layer thickness decreases, enhancing the mass transport of reactive species (e.g., H^+ or O_2) to the electrode surface. This leads to:

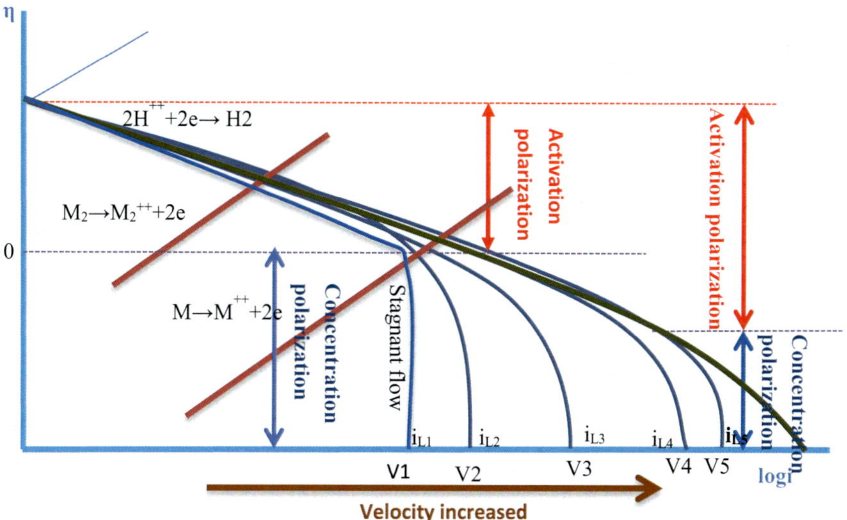

Fig. 7.5 The effect of velocity on the limiting current, activation polarization, and concentration polarization

- A higher limiting current (i_{lim})
- A reduction in concentration polarization
- Greater contribution from activation polarization, especially in the early stages of velocity increase

Eventually, at sufficiently high velocities, the system may become entirely activation-controlled, where the reaction rate depends mainly on electrode kinetics rather than mass transport.

At location (1) in Fig. 7.5, the flow is stagnant, meaning the fluid velocity is zero. Under this condition, the limiting current equals the corrosion current, the anodic current, and the cathodic current:

$$i_{\text{lim}} = i_{\text{corr}} = i_a = i_c$$

Because mass transport is severely restricted, the electroactive species at the electrode surface are completely consumed, and no replenishment occurs from the bulk solution. As a result, the system experiences infinite concentration polarization, indicating a fully diffusion-limited corrosion process.

As mentioned earlier, when the fluid velocity increases, the limiting current (i_{lim}) also increases due to enhanced mass transport of reactive species to the electrode surface. As a result, the system experiences a reduction in concentration polarization.

At the same time, activation polarization becomes more dominant, as the electrochemical reaction rate is now increasingly controlled by electrode kinetics rather than diffusion. In this regime, the system transitions from being mass-transport-limited to being activation-controlled, particularly at higher flow velocities.

Eventually, at a sufficiently high flow velocity greater than V_5, the cathodic concentration polarization disappears entirely. This condition is illustrated by the green line in Fig. 7.5, which shows a purely activation-controlled cathodic response.

At this stage, mass transport is no longer a limiting factor, and the corrosion process is governed solely by the kinetics of the electrochemical reaction at the electrode surface. This marks the transition to a regime where the system's behavior is dominated by activation polarization, and the overall corrosion rate depends on reaction rates rather than species diffusion.

As the fluid velocity increases, the system transitions from diffusion-limited to activation-controlled behavior. Under these conditions, the anodic polarization curve of the metal intersects with the hydrogen cathodic reaction curve within the activation polarization region.

In this case, the limiting current ($i_{\mathbf{lim}}$) is greater than the corrosion current ($i_{\mathbf{corr}}$), meaning that mass transport no longer restricts the reaction rate. Instead, the rate of corrosion is governed by the electrode kinetics, that is, the ability of the electrochemical reaction to proceed at the electrode surface.

Thus, when operating in the activation-controlled region, the corrosion current is less than the limiting current, and concentration polarization becomes negligible.

$$i_{L1} < i_{L2} < i_{L3} < \ldots i_{Ln}$$

$$i_{corr1} < i_{corr2} < i_{corr3} < \ldots i_{corr\,n}$$

$$i_{L1} = i_{corr1}$$

$$i_{L2} \neq i_{corr2}$$

Figure 7.6 illustrates the effect of fluid velocity on the corrosion rate for different metals. As shown, flow velocity does not continuously increase the corrosion rate beyond a certain threshold.

For metal M_2, the corrosion rate initially increases gradually with velocity due to enhanced mass transport. However, beyond a specific velocity, the rate levels off and becomes constant, indicating that the system has transitioned to an activation-controlled regime where further increases in velocity have no effect. In this case, velocity has a limiting influence on corrosion rate.

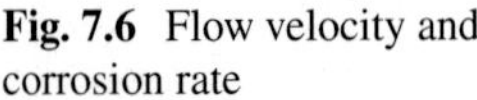
Fig. 7.6 Flow velocity and corrosion rate

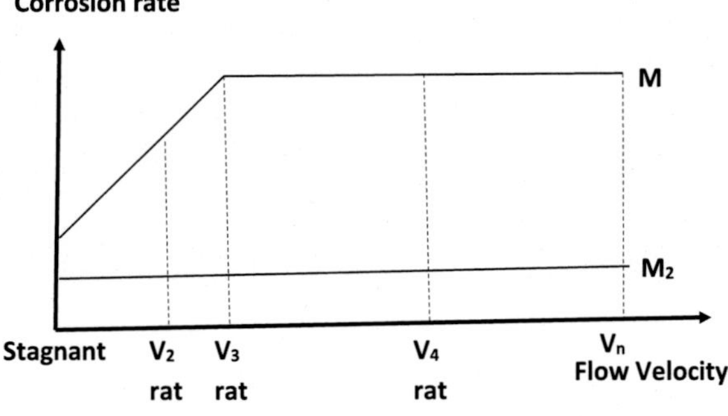

For metal M_2, the corrosion rate remains constant at all velocities. This behavior is attributed to the dominance of activation polarization and the absence of concentration polarization, meaning the corrosion process is controlled entirely by surface reaction kinetics, not by diffusion.

These trends, shown in Figs. 7.5 and 7.6, highlight the difference between diffusion-limited and activation-controlled corrosion systems and the importance of understanding polarization behavior in predicting the effect of flow.

Chapter 8
Passivation

Corrosion resistance in many metals arises from the formation of a protective surface film, such as a sulfate, oxide, or hydroxide, under oxidizing conditions accompanied by high anodic polarization. This passive film acts as a barrier, significantly reducing the rate of further metal dissolution.

Faraday's experiments provide a foundational understanding of this behavior. His work demonstrated that, beyond a certain applied potential, metals like iron, chromium, and aluminum exhibit a sudden drop in current density, indicating the formation of a passivating layer. This phenomenon, known as passivation, explains why some metals become resistant to corrosion despite being thermodynamically unstable in their environment.

Faraday's Experiment
A piece of iron is immersed in three separate beakers to observe the effect of acid concentration on corrosion behavior:

- The first beaker contains concentrated nitric acid (HNO_3).
- The second and third beakers contain dilute nitric acid.

This experiment demonstrates the phenomenon of passivation. In concentrated nitric acid, the iron quickly forms a protective oxide film, which inhibits further corrosion, placing the metal in a passive state. This occurs due to the strong oxidizing nature of concentrated HNO_3, which promotes rapid surface film formation.

In contrast, in dilute nitric acid, the oxidizing power is insufficient to form a stable passive layer. As a result, the iron remains active, and corrosion continues due to ongoing metal dissolution.

A piece of iron is first immersed in a beaker containing concentrated nitric acid. After a short period, bubbles begin to form at the surface of the iron. These bubbles are identified as nitric oxide (NO_2) gas, produced by the reaction between the acid and the metal. Simultaneously, a layer of iron oxide forms around the surface of the iron.

M. M. Ghatus, *Fundamentals of Corrosion Science and Engineering*,
https://doi.org/10.1007/978-3-032-13138-6_8

This oxide film acts as a barrier, preventing further contact between the nitric acid and the iron surface. As a result, oxidation ceases, and the metal becomes passivated.

The same iron piece is then removed and transferred into a second beaker containing dilute nitric acid. Due to the presence of the previously formed oxide film, no reaction occurs. The passive layer prevents any further interaction between the iron and the dilute acid.

Finally, the iron piece is removed again, scratched to damage the passive film, and immersed in a third beaker also containing dilute nitric acid. At the scratched area, the protective film is broken, and a reaction resumes. Nitric oxide (NO_2) gas evolves at the exposed metal surface, confirming that the passivation was locally disrupted, and corrosion is re-initiated at the exposed area.

The anodic and cathodic reactions of the Faraday's experiment are:

$$HNo_3 + H^+ + e \rightarrow NO_2 + H_2O \text{ Cathodic reaction}$$

$$Fe \rightarrow Fe^{++} + 2e \text{ Anodic reaction}$$

Once again, high corrosion resistance is attributed to significant anodic polarization, which corresponds to a large anodic overvoltage. This condition reflects a high positive potential relative to the metal's equilibrium potential.

In other words, when a metal is subjected to large anodic polarization, it reaches a potential above zero (vs. SHE), where the formation of a stable passive film becomes thermodynamically favorable. This positive shift in potential hinders further metal dissolution and is a key factor in the passivation behavior observed in metals like iron, chromium, and aluminum.

A large anodic polarization not only reflects a positive potential but also corresponds to a high anodic current, indicating an increased rate of electron release due to metal oxidation. In a polarization test (as shown in Fig. 8.1), the potential is intentionally shifted in the positive direction to evaluate the metal's behavior under oxidizing conditions.

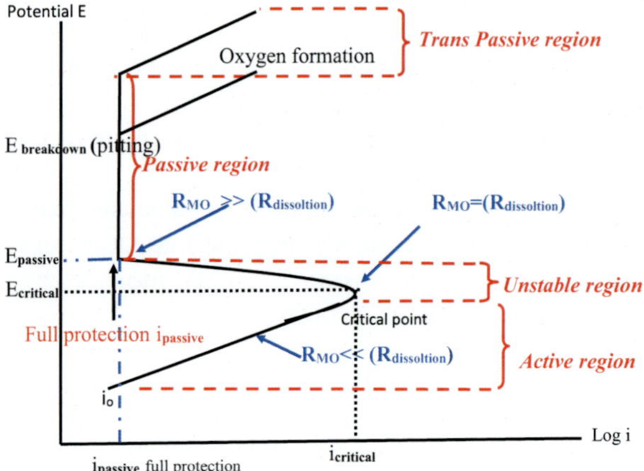

Fig. 8.1 Passivation in logi-E diagram

For example, when a metal is immersed in water, beyond a certain potential and at a specific pH level, the conditions may become favorable for the formation of metal oxides or hydroxides. This is predicted by the Pourbaix diagram, which outlines the thermodynamic stability of corrosion products based on potential and pH.

This process can be understood in two key steps:

(a) Electrochemical Activation:

As the potential increases, the metal enters the active region, where it begins to oxidize. Anodic current increases, and metal ions (e.g., M^{2+}) are released into the solution.

(b) Passive Film Formation:

At higher potentials, under suitable pH conditions, the metal ions react with water or hydroxide ions, forming a surface layer of metal oxide or hydroxide. This layer acts as a barrier, significantly reducing further anodic current and transitioning the metal into the passive region.

- Metal dissolution rate ($R_{\mathbf{dissoltion}}$).
- Metal oxide formation rate (R_{MO}).

$$M + H_2O \rightarrow MO + 2H^+ + 2e$$

When:

$$R_{MO} << R_{\mathbf{dissoltion}} \qquad \text{Active region}$$

An increase in $R_{MO,}$ the resistance due to the metal oxide film, indicates greater surface coverage of the metal by the oxide layer. As this protective film thickens or becomes more stable, it acts as a barrier to ion transport and electron exchange at the metal/electrolyte interface.

Consequently, the rate of metal dissolution ($R_{\mathbf{dissolution}}$) decreases as the oxide layer impedes the anodic reaction. This relationship reflects the passivation process, where the formation of a stable metal oxide reduces the corrosion rate significantly.

Once it reaches some point at:

$$R_{MO} = R_{\mathbf{dissoltion}} \qquad \text{Unstable region}$$

Gradually increasing the RMO will decrease the rate of metal dissolution.

$$R_{MO} >> R_{\mathbf{dissoltion}} \qquad \text{Passivation region}$$

When the metal oxide layer fully covers the metal surface, the electrochemical reaction is significantly suppressed. At this stage, the current measured is known as the passivation current ($i_{\mathbf{passive}}$), which indicates a state of full protection.

The value of $i_{\mathbf{passive}}$ is typically very small, especially when plotted on a logarithmic scale, reflecting an extremely low rate of metal dissolution. This low current confirms the effectiveness of the passive film in preventing further corrosion, as the anodic reaction becomes nearly negligible.

Once again, sizeable anodic polarization leads to forming a passive region, but what will happen to the passive region if the potential continues to increase (anodic polarization)?

As shown in the Pourbaix diagram, there exists a specific region that represents the stability zone of water (H_2O). Within this zone, water remains thermodynamically stable; however, outside of it, at certain pH and potential values, water can undergo decomposition reactions, resulting in the formation of oxygen or hydrogen gas.

In a neutral or basic medium, oxygen evolution becomes possible at higher anodic potentials, following the reaction:

$$O2 + 4H^{+} + 4e^{-} \leftrightarrow 2H2O(\text{acidic medium})$$

or in a more relevant form for neutral/basic media:

$$O2 + 2H2O + 4e^{-} \rightarrow 4OH^{-}$$

This reaction defines the upper boundary of water stability in the Pourbaix diagram and sets a limit for safe anodic polarization to prevent oxygen gas evolution from water decomposition.

The passive region, where the metal surface is protected by a stable oxide or hydroxide film, can be compromised in the presence of aggressive anions, particularly chloride ions (Cl^-). These small, highly mobile ions can penetrate or disrupt the passive film, leading to localized breakdown and initiating pitting corrosion.

The breakdown of the passive layer in the presence of chloride ions can be represented by the following reaction:

$$\text{Me}^{n+} + nCl^{-} \rightarrow \text{MeCl}_{n}$$

This reaction illustrates how metal cations (Me^{n+}) formed at weak points in the passive film react with chloride ions to form soluble metal-chloride complexes, which further destabilize the protective film and accelerate localized dissolution.

In the case of chromium, the passive film that forms on the surface is primarily composed of chromium(III) oxide (Cr_2O_3). This protective layer provides excellent resistance to further corrosion by limiting ion transport and isolating the metal from the environment.

During passivation, chromium ions (Cr^{3+}) are generated at the metal surface as a result of anodic dissolution

$$2\text{Cr}^{+3} + 7\text{H}_2\text{O} \rightarrow \text{Cr}_2\text{O}_7\ (\text{soluble}) + 14\text{H}^{+} + 6\text{e}$$

This transformation indicates a shift from passive film stability (Cr_2O_3) to the formation of soluble chromate species, which may signal the onset of film degradation or transpassivity, especially under highly oxidizing conditions.

Aluminum also forms a passive film composed of aluminum oxide (Al_2O_3). However, unlike some other passive films, the Al_2O_3 layer is often porous, primarily due to the potential gradient that develops between the metal surface and the metal oxide interface.

As the applied potential increases, the electric field across the oxide layer becomes stronger. This increased potential gradient enhances the migration of Al^{3+} ions from the metal substrate toward the electrolyte. At high potentials, the rate of aluminum ion migration exceeds the rate of oxide formation, preventing the development of a compact, protective layer. Instead, the oxide grows in a porous structure.

This phenomenon is known as anodizing, a process intentionally used to form a controlled, porous Al_2O_3 film on aluminum surfaces. Anodizing is widely used in industry to improve corrosion resistance, paint adhesion, and surface hardness.

The Conclusion of Transpassivity Is Due To
Key phenomena that occur under high anodic polarization include:

- **Oxygen Evolution:**

At high positive potentials, water undergoes oxidation, leading to the formation of oxygen gas (O_2), particularly in neutral or basic environments:

$$2H_2O \rightarrow O_2 + 4H^+ + 4e^-$$

- **Chloride-Induced Film Breakdown:**

Chloride ions (Cl^-) can penetrate or disrupt passive films, especially on stainless steels or aluminum, initiating localized corrosion such as pitting or crevice corrosion.

- **Anodizing (in Aluminum):**

In the case of aluminum, high anodic potentials lead to the formation of a porous Al_2O_3 layer, as the migration rate of Al^{3+} ions exceeds the oxide formation rate. This controlled oxide growth process is known as anodizing.

- **Chromium Oxide Dissolution and Chromate Formation:**

For chromium, the passive Cr_2O_3 layer may dissolve at high potentials. The released Cr^{3+} ions react with water and undergo further oxidation to form chromate ions ($Cr_2O_7^{2-}$):

$$2Cr^{3+} + 7H_2O \rightarrow Cr_2O_7^{2-} + 14H^+ + 6e^-$$

8.1 Passivation and Mixed Potential Theory

The interplay between the cathodic and anodic reactions is illustrated in Fig. 8.2, which shows the potential–log current density (E–log i) relationship. The cathodic reaction influences the potential of the metal undergoing dissolution, potentially shifting it into the:

- Active region (metal actively corroding).
- Passive region (metal protected by a surface film).
- Unstable passive region (film breakdown or localized corrosion).

If the metal is passivated, it means its potential has increased sufficiently to form a protective oxide or hydroxide layer, a process often referred to as anodizing in controlled environments. In the curve shown in Fig. 8.2, the metal transitions from an active region to an unstable zone, and eventually into the passive region, indicating the formation of a stable protective layer at higher anodic polarization.

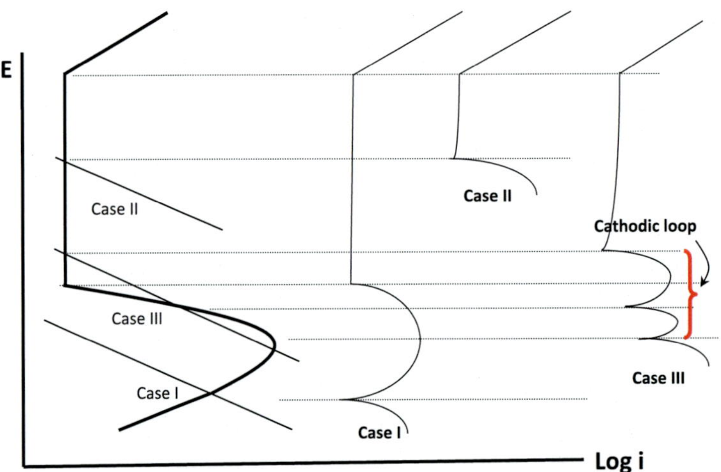

Fig. 8.2 Passivation and mixed potential theory

In Case III, the system exhibits unstable passivation. The cathodic reaction curve intersects the anodic polarization curve at three distinct points, corresponding to the passive, unstable, and active zones. This behavior results in the formation of a cathodic loop, which defines a region of electrochemical instability. Within this zone, the system can fluctuate between localized film formation and breakdown, making it highly susceptible to pitting or crevice corrosion.

As an example, titanium (Ti), a highly active metal, can be intentionally alloyed with noble metals such as palladium (Pd) or platinum (Pt) to enhance its corrosion resistance. When Pd or Pt is homogeneously mixed with Ti, the resulting alloy benefits from both the passivating behavior of Ti and the catalytic cathodic properties of the noble metal.

One of the primary advantages of titanium in such alloys is its ability to achieve passivation readily due to cathodic polarization, particularly in corrosive environments.

In Case I (Fig. 8.2), when a Ti-Pt alloy is immersed in an air-free acidic solution, the dominant cathodic reaction is hydrogen evolution, driven by the acidic medium. On the Ti surface, the anodic polarization curve follows four distinct stages, as shown in Fig. 8.3:

1- Active region—continuous metal dissolution.
2- Unstable region—transition between activity and passivation.
3- Passive region—formation of a stable TiO_2 film.
4- Transpassive region—breakdown or excessive oxidation of the passive layer.

During the initial immersion, both Ti and Pt may undergo uniform dissolution. However, due to Pt′s higher standard reduction potential, it is more noble than Ti and tends to be redeposited onto the surface through a dealloying phenomenon.

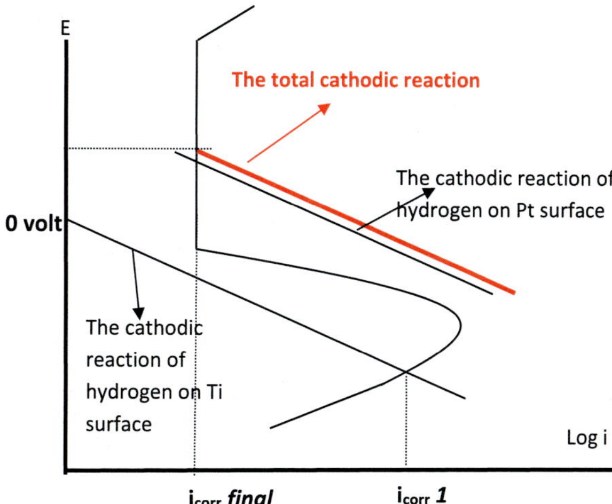

Fig. 8.3 Ti-Pt in air-free acid

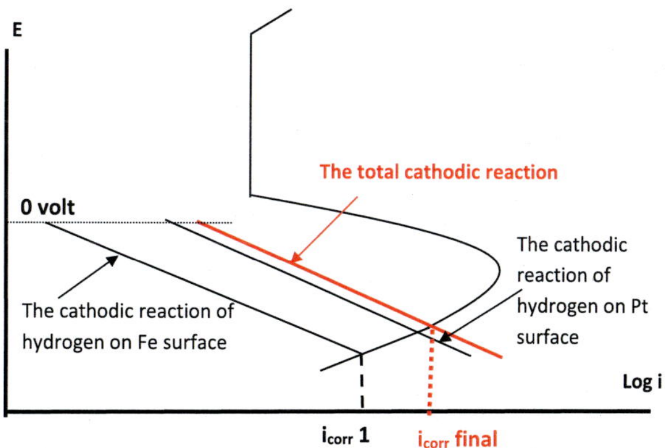

Fig. 8.4 Fe-Pt in air-free acid

As a result, the surface becomes enriched in Pt, which catalyzes the hydrogen reduction reaction more effectively than Ti. This change leads to a significant decrease in corrosion current, transitioning from i_{corr1} (initial corrosion current) to $i_{corr\ final}$, as illustrated in Case II of Fig. 8.2.

This behavior exemplifies how strategic alloying with noble metals like Pt or Pd can shift the corrosion mechanism, enhance cathodic kinetics, and stabilize the passive state, ultimately improving the alloy's corrosion resistance in aggressive environments.

When platinum (Pt) is alloyed with iron (Fe) and the alloy is exposed to an air-free acidic environment, the dominant cathodic reaction is the hydrogen reduction reaction. However, in this case, the addition of Pt has a detrimental effect on corrosion behavior.

As shown in Fig. 8.4, the cathodic polarization curve for hydrogen reduction on the Fe–Pt alloy surface is significantly shifted away from the passivation region of iron. This is because Pt, being highly noble, dramatically enhances the cathodic kinetics of hydrogen evolution, resulting in a much higher cathodic current at any given potential.

As a consequence, the corrosion rate increases significantly:

$$i_{\text{corr final}} \gg i_{\text{corr 1}}$$

This increase occurs because the iron surface can no longer reach or maintain the passive region, as the cathodic demand (due to enhanced hydrogen evolution on Pt sites) prevents the stabilization of a passive film.

Thus, unlike the Ti–Pt system, where Pt promotes passivation, the addition of Pt to Fe results in a negative effect by suppressing passivation and accelerating uniform corrosion. The Fe–Pt alloy, under these conditions, remains active, and passivation is not achieved.

8.2 The Velocity Effect on the Corrosion Rate of Passivated Metal

To understand the effect of fluid velocity on passivated metals, it is necessary to examine the combined impact of concentration polarization and passivation behavior.

Figure 8.5 illustrates the relationship between electrode potential and logarithmic current density (log i) for a specific passivated metal (M). The cathodic reaction in this system consists of two key components:

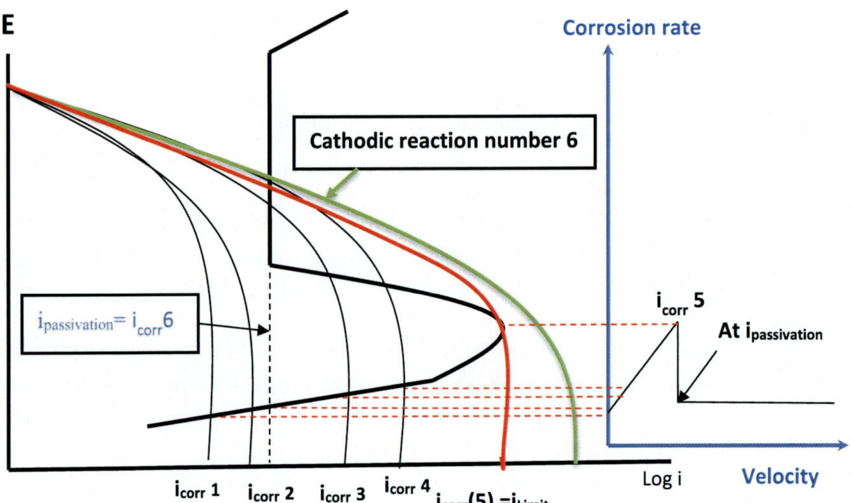

Fig. 8.5 Shows the effect of velocity on passivated material

- Activation polarization (kinetic control).
- Concentration polarization (mass transport control).

At progressively increasing fluid velocities, the cathodic polarization curve shifts due to enhanced mass transport:

- At velocity one, the cathodic curve intersects the anodic curve at i_{corr}.
- At velocity two, the corrosion current is i_{corr2}.
- Similarly, i_{corr3} and i_{corr4} are reached at higher velocities.

At velocity five, the corrosion current (i_{corr}) reaches the limiting current (i_{lim}), representing the maximum current achievable under diffusion-limited conditions. This corresponds to location five, where the velocity is higher than at location one.

However, in cathodic reaction curve number 6, the intersection occurs within the passivation region of the anodic curve. In this case, the corrosion current suddenly drops to the passivation current ($i_{passive}$). This indicates that further increases in velocity no longer affect the corrosion rate, the metal is fully passivated, and a stable oxide layer protects the surface.

Additionally, Fig. 8.5 shows the overall relationship between corrosion rate and fluid velocity:

- At low velocities, the corrosion rate increases with velocity due to enhanced cathodic kinetics.
- Once the limiting current equals the corrosion current, a transition occurs.
- When the system enters the passivation region, the corrosion current abruptly drops to $i_{passive}$.
- Beyond this point, the corrosion rate remains constant, regardless of further velocity increases.

It is also important to note that cathodic curves 3 and 4 intersect the anodic polarization curve at three points: in the active, unstable, and passivation regions. However, the intersections in the unstable and passive zones are not meaningful unless a passive film has already formed. The system will follow the active dissolution path until conditions allow for the transition into stable passivation.

8.3 The Effect of Oxidizer Concentration on the Active-Passive Metal

Figure 8.6 demonstrates the effect of oxidizer concentration on an active-passive metal system. The cathodic reduction reaction is influenced directly by the oxidizer's concentration; as the concentration increases, the cathodic potential shifts positively. This is illustrated by the series of cathodic curves labeled from 1 to 9.

- From cathodic reactions 1 to 5, the corrosion current gradually increases due to the enhanced cathodic kinetics with increasing oxidizer concentration.

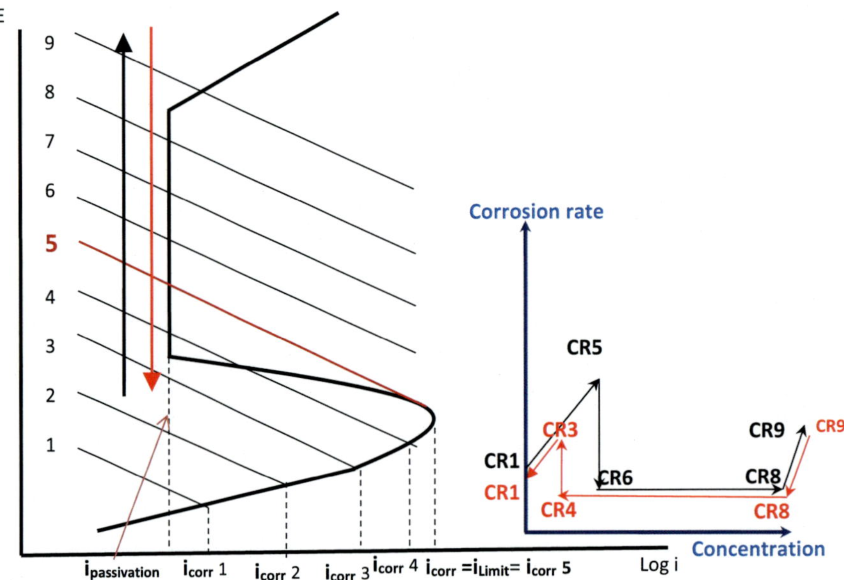

Fig. 8.6 The effect of oxidizer concentration on the active–passive metal

- At cathodic curve 5, the limiting current equals the corrosion current ($i_{lim} = i_{corr}$), marking the maximum corrosion rate for that range.
- At cathodic curve 6, the reaction intersects only the passive region of the anodic polarization curve. This results in a sudden drop in corrosion current to the passivation current ($i_{passive}$).
- From curves 6 to 8, the corrosion current remains constant, as the system is fully passivated. Further increases in oxidizer concentration no longer increase corrosion.
- At cathodic reaction 9, the intersection occurs within the transpassive region, where the passive film begins to break down. As a result, the corrosion rate increases again.

Now Consider the Reverse Scenario

If the system begins at cathodic reaction 9 and the oxidizer concentration is decreased, the metal is already in a passivated state. The corrosion path follows the passive branch (red curve) until the concentration drops to the level corresponding to cathodic curve 4.

- At curve 4, the cathodic reaction intersects three anodic regions: passive, unstable, and active. Although the system is still passivated, it becomes susceptible to film instability, especially if mechanical damage or local chemistry change occurs.
- At curve 3, the cathodic curve intersects only the active region. This indicates that the passive film has broken down, and active corrosion resumes.

- As oxidizer concentration continues to decrease (moving to curves 2 and 1), the corrosion rate gradually declines due to reduced cathodic driving force, but the system remains in the active corrosion state.

This behavior highlights the critical role of oxidizer concentration in sustaining or disrupting passivity. The system exhibits hysteresis once passivity is lost; restoring it may require significantly higher oxidizer concentrations or external polarization.

Chapter 9
Different Corrosion Protection Mechanisms, Electrochemical Ways of Protection, and Cathodic Protection

Protection against corrosion is essential for extending the service life of equipment and minimizing safety and operational risks. The primary goal is to drastically reduce the corrosion rate, thereby improving the reliability and performance of metallic components in corrosive environments.

Figure 9.1 presents a comprehensive overview of the various corrosion protection methods, including:

- Protective coatings.
- Material selection and alloying.
- Environmental modification.
- Corrosion inhibitors.
- Cathodic protection.
- Anodic protection.
- Design improvements.

Among these strategies, this book places particular emphasis on cathodic protection and anodic protection due to their scientific basis, industrial significance, and effectiveness in controlling electrochemical corrosion. These methods are explored in detail in the following sections.

9.1 Cathodic Protection (CP)

Basics of Cathodic Protection

The fundamental principle of cathodic protection is to convert the metal structure at risk of corrosion into a cathode, thereby suppressing its anodic dissolution. This is achieved through the following core steps:

1. Electron Supply to the Protected Metal:

M. M. Ghatus, *Fundamentals of Corrosion Science and Engineering*,
https://doi.org/10.1007/978-3-032-13138-6_9

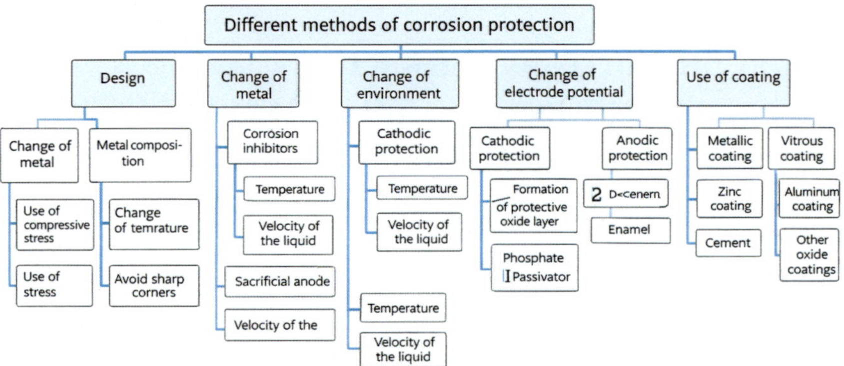

Fig. 9.1 Methods of corrosion protection

An external source of electrons is provided, typically through a negative direct current (DC). This external cathodic current (i_c) supplies electrons to the metal surface, preventing oxidation (metal dissolution).

2. Electrode Configuration and Current Flow:

The metal to be protected is connected to the negative terminal of a DC power supply, designating it as the cathode.

An auxiliary electrode (AE), often made of an inert or consumable anode material, is connected to the positive terminal.

- Cathodic reaction (e.g., hydrogen evolution or oxygen reduction) occurs on the protected metal surface.
- Anodic reaction (e.g., metal dissolution or oxygen evolution) occurs at the auxiliary electrode (AE).

As shown in Fig. 9.2, this arrangement ensures that the protected metal remains cathodic, effectively eliminating its corrosion by shifting the electrochemical reactions away from its surface.

Cathodic protection (CP) is widely applied to safeguard steel structures exposed to corrosive environments, such as buried pipelines, offshore platforms, and marine vessels. It is particularly effective in situations where there is intimate contact between the metal surface and the electrolyte, such as soil, water, or moist environments.

However, cathodic protection is only effective in regions where electrical continuity and electrolyte access are maintained. Any part of the structure that lacks intimate contact with the electrolyte, such as areas insulated by air gaps or shielding, cannot be protected using this method.

CP is especially beneficial for coated structures, such as painted or epoxy-lined pipelines. In these systems, cathodic protection works synergistically with the coating:

- The coating reduces the overall current demand.

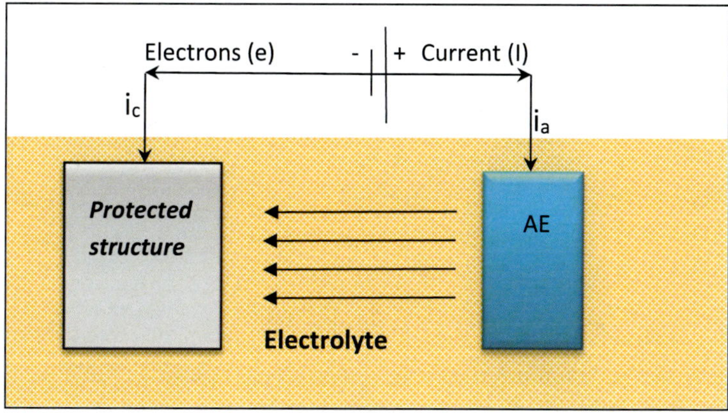

Fig. 9.2 Cathodic protection

- CP protects the uncoated spots, coating holidays, or damaged areas, where localized corrosion would otherwise occur.

Additionally, cathodic protection is used effectively in shielded configurations, such as the inner members of a bundle of pipes, where direct coating or maintenance is challenging. In these cases, CP provides general and localized protection by maintaining the electrochemical potential below the corrosion threshold.

Limitations of Cathodic Protection

- Cathodic protection (CP) cannot be effectively applied in the following scenarios:

Above the waterline: No continuous electrolyte contact exists, so current flow and electrochemical reactions cannot be maintained.

- Non-conductive media: In environments such as dry gases or certain plastics, the absence of ionic conductivity prevents the formation of an electrochemical cell.

Extremely corrosive environments: In such cases, a significantly higher current is required to achieve protection, which may render the system economically or technically impractical due to excessive power consumption.

Relationship Between Potential and Current in CP (Fig. 9.3).

Figure 9.3 illustrates the relationship between electrode potential and current during cathodic protection. When electrons are supplied to the protected metal (through an external power source), the cathodic current ($|i_c|$) increases, and the anodic current ($|i_a|$) decreases.

This shift disrupts the electrochemical equilibrium, creating a net current:

$$i_{\text{applied}} = |i_c| - |i_a|$$

This applied current is the current that must be supplied externally to achieve protection. The magnitude of i_{applied} is directly proportional to the system's energy (or power) requirement.

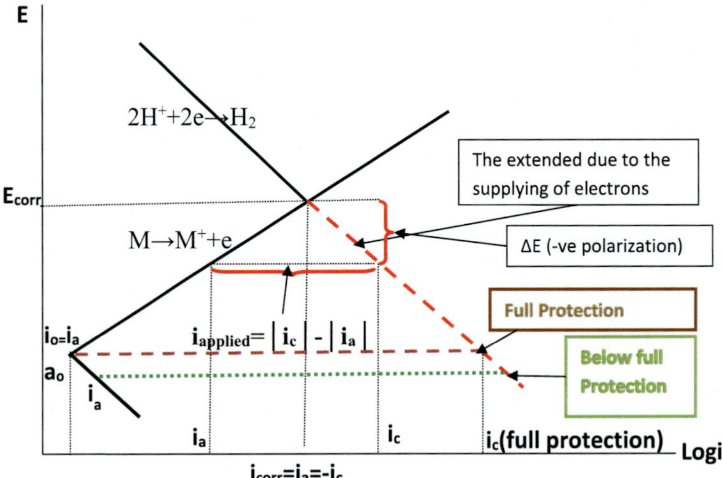

Fig. 9.3 Cathodic protection in E-logi diagram

Therefore, minimizing the difference between $|i_c|$ and $|i_a|$, i.e., optimizing the protection currentreduces the operational cost and improves the system's efficiency.

To achieve complete protection of a metal surface through cathodic protection, the cathodic current density must be sufficient to suppress the anodic reaction entirely. This occurs when the exchange current density (i_o) on the metal surface equals the anodic current (i_A):

$$i_o = i_A$$

However, achieving this condition requires a very high applied current, which is typically impractical and energy-intensive, making it undesirable in real applications.

Cathodic Overprotection and Cathodic Corrosion

If the cathodic current (i_c) applied to the metal exceeds the level required for complete protection, it can lead to undesirable side effects. One of these is cathodic corrosion, a phenomenon that occurs when:

- The applied cathodic current surpasses the optimal level.
- The anodic current on the metal still exceeds the natural current density of the metal.
- Hydrogen evolution or embrittlement occurs due to excess polarization.

As shown in Fig. 9.3, this overpolarization causes the metal to corrode despite being cathodically polarized, typically due to unintended reactions, such as:

- Hydride formation in metals like titanium.
- Hydrogen embrittlement in high-strength steels.

This condition, known as cathodic corrosion, underscores the importance of maintaining an optimal protection current, rather than maximizing it blindly.

Types of Cathodic Protection
There are two main systems used in practice:

1. Sacrificial Anode Cathodic Protection (SACP):
   A more active metal (e.g., zinc, magnesium, and aluminum) is connected to the protected structure.
   The anode corrodes preferentially, supplying electrons to the protected metal.
   No external power source is required
2. Impressed Current Cathodic Protection (ICCP):
   - Uses an external DC power supply to drive current from an inert or semi-inert anode (e.g., graphite, MMO, platinum).
   - Suitable for larger structures or high-resistance environments.
   - Allows controlled and adjustable current output.

9.1.1 Sacrificial Anode

A sacrificial anode is a fundamental device used in cathodic protection systems to protect a metallic structure, such as pipelines, tanks, or marine structures, from corrosion by deliberately corroding a more active metal. This approach is based on the principles of the electrochemical series, where:

The metal to be protected must be nobler (less active) than the sacrificed anode metal.

As illustrated in Fig. 9.4, a steel pipeline buried in soil is connected to a sacrificial anode made of an active metal such as zinc or magnesium. The active metal corrodes preferentially and supplies electrons to the steel pipeline through a connecting wire. The current flows through the electrolyte (soil), completing the circuit.

This electron flow shifts the potential of the steel pipeline in the cathodic direction, thereby suppressing its corrosion.

Uneven Current Distribution and the Role of Backfill
In practice, the conductivity of the electrolyte (soil) may vary, causing non-uniform current distribution around the anode's surface. As a result:

The current may leave the anode at localized spots

- Leading to accelerated, uneven corrosion on the anode surface.
- Which in turn reduces the effective lifespan of the sacrificial anode.

To address this issue, a backfill material, typically made of bentonite, sodium chloride, gypsum, or coke, is placed around the anode. The purpose of the backfill is to:

- Create uniform conductivity around the anode surface.

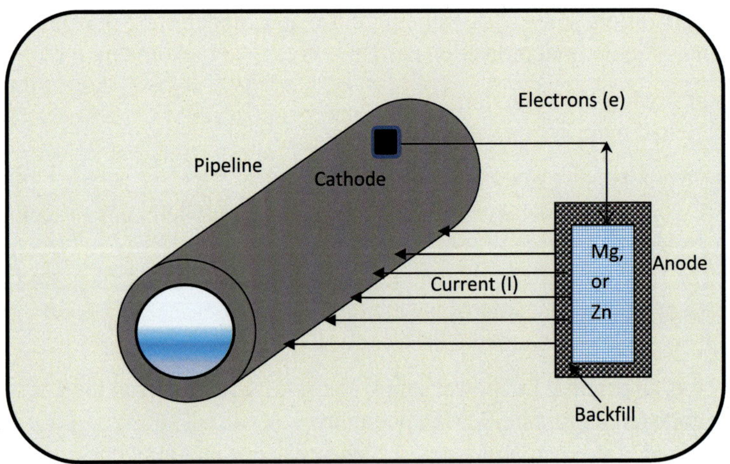

Fig. 9.4 Cathodic protection by sacrificial anode

- Prevent localized corrosion and promote uniform corrosion.
- Enhance anode efficiency and maximize its service life.
- Maintain stable electrical contact with the surrounding electrolyte.

9.1.2 Characteristics of the Sacrificial Anode Method

The sacrificial anode cathodic protection method operates on electrochemical principles and is widely valued for its simplicity and self-sufficiency. Its effectiveness depends on several key characteristics:

1. Sufficient Potential Difference:

   There must be an adequate electrochemical potential difference between the sacrificial anode (more active metal) and the structure to be protected (cathode). This difference drives the flow of electrons from the anode to the cathode, enabling effective protection.
2. High Amp-Hour Efficiency (Electrical Energy Output):

   The anode material must have a high current capacity per unit mass, typically measured in amp-hours per kilogram (Ah/kg). This value determines how long a given mass of sacrificial anode can provide protection. A high amp-hour rating ensures a longer service life of the anode.
3. No External Power Source Required:

   The system operates without the need for an external electrical power supply. The galvanic coupling between the anode and cathode creates a natural current flow, making the sacrificial anode method self-powered, simple, and especially effective in remote or submerged environments (Table 9.1).

Table 9.1 Typical environment and operating characteristics of most common anodes (AE)

Anode material	Typical environment	Typical environment resistivity (Ω·m)	Operating resistivity (Ω·m)	Approximate consumption rate (kg/A·year)
Zinc	All aqueous environments	0.2–15	Controlled by circuit resistance	12
Aluminum	Potable water	0.2–4	Controlled by circuit resistance	3.5
Magnesium	Seawater or potable water	5–75	Controlled by circuit resistance	7
Platinized titanium	Sea or potable water	0.2–75	100–1000	1×10^{-5}
Silicon iron	All aqueous environments	0.2–75	5–40	0.3–1
Magnetite	All aqueous environments	0.2–75	3–60	<0.1
Scrap steel	All aqueous environments	10–200	0.1–1	10
Mixed metal oxide	All aqueous environments	0.2–200	60–750	1×10^{-6}

9.1.3 Impressed Current (ICCP)

The Impressed Current Cathodic Protection (ICCP) method utilizes an external power source to protect metal structures from corrosion. As illustrated in Fig. 9.5, the setup consists of the following:

1. Electrochemical Configuration:
   - The protected metal is connected to the negative terminal of the power source, while the auxiliary electrode (AE) is connected to the positive terminal.
   - The AE serves as a current delivery point, completing the electrochemical circuit, thereby driving the cathodic reaction on the metal to be protected.
2. Backfill and Current Distribution:
   - Similar to the sacrificial anode method, the AE is surrounded by a backfill material to ensure uniform current distribution across the protected metal surface.
   - In certain cases, more than one AE is used to enhance the current distribution and ensure that all surfaces of the metal are effectively protected.
3. Anode Materials:

   Anodes in the ICCP system typically consist of materials such as:

   - Graphite.
   - High Silicon Iron.
   - Platinumized Titanium.

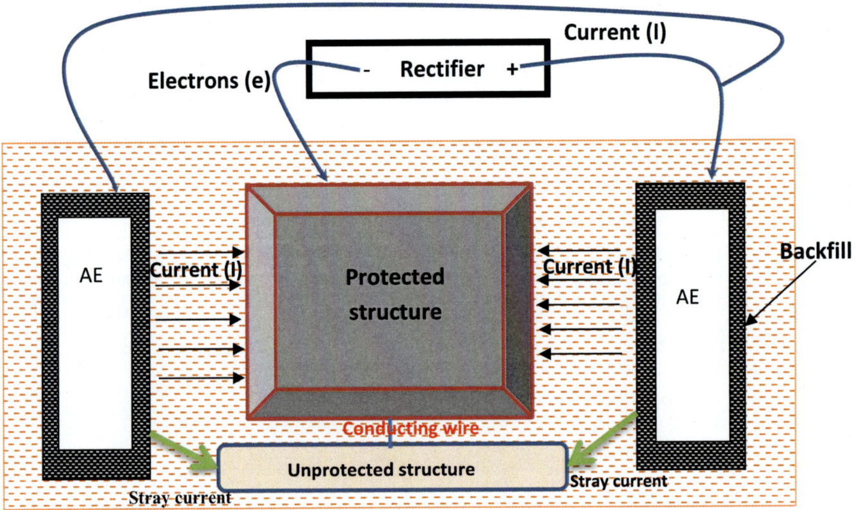

Fig. 9.5 Cathodic protection by impressed current

- Platinized Niobium (Tantalum).
- Magnetite Oxides.
- Mixed Metal Oxides (MMO).

4. Stray Current Problem:

    - A common issue in ICCP systems is the occurrence of stray current, which occurs when current flows out of the AE into the surrounding electrolyte or soil, affecting nearby unprotected structures.
    - Stray currents can cause severe corrosion at uncoated or anodic areas on unprotected structures. These areas, being small compared to the overall unprotected structure, are particularly vulnerable.
    - To mitigate the stray current issue, unprotected structures can be connected to the protected metal through a conductive wire, integrating them into the ICCP system and ensuring uniform protection across all surfaces.

Advantages of Impressed Current Cathodic Protection (ICCP)

The ICCP system offers several important benefits, making it a widely used method in both industrial and infrastructure applications:

1. Versatility:
    ICCP can be applied to a wide range of structures, including large, complex geometries such as ships, pipelines, storage tanks, and offshore platforms.
2. Effectiveness in High-Resistance Environments:
    Unlike sacrificial anode systems, ICCP is effective even in high-resistivity media, such as dry soil, concrete, or high-resistance water, due to its adjustable power input.

3. Coating Quality Monitoring:
   ICCP allows for real-time monitoring of protective coating integrity. Areas with coating damage or degradation demand more current, revealing potential failures through system diagnostics.

Disadvantages of Impressed Current Cathodic Protection (ICCP)
Despite its advantages, ICCP has some notable limitations:

1. High Installation and Operating Cost:
   ICCP systems require a power source, control units, and durable anode materials, making them more expensive than sacrificial anode systems in terms of both initial setup and maintenance.
2. Stray Current Issues:
   ICCP may cause stray current corrosion in adjacent, unprotected structures. This occurs when current unintentionally flows through neighboring metallic objects, causing accelerated anodic dissolution at unintended sites.

9.2 Anodic Protection

Anodic protection is based on the tendency of certain metals and alloys to passivate—that is, to form a stable, protective oxide film when subjected to controlled anodic polarization. This method is effective only for materials that exhibit passivation behavior, such as stainless steels, chromium alloys, and titanium.

Anodic protection relies on shifting the potential of the metal from its free corrosion potential into the passive region of its polarization curve. Once this potential is reached, the anodic current drops significantly, indicating the formation of a protective surface film and a corresponding reduction in corrosion rate.

This technique is not suitable for active metals that do not exhibit passivation (e.g., carbon steel in non-oxidizing acid solutions), as these materials continue to corrode actively regardless of anodic polarization.

In terms of electrochemical behavior:

- During anodic polarization, the anodic current decreases as the system moves into the passive region, while the cathodic current remains low.
- Conversely, in cathodic polarization, the cathodic current increases, and the anodic current is suppressed.

This inverse relationship between anodic and cathodic currents underscores the importance of maintaining precise control of potential in anodic protection systems to ensure that the metal remains within the stable passive zone. This process is referred to as anodic polarization.

In Fig. 9.6, the electrode potential of the metal is increased by an amount ΔE, shifting the system from the free corrosion potential into the anodic polarization region. In this region, the anodic current (i_a) becomes greater than the cathodic current (i_c), indicating that the metal is being actively driven toward passivation.

As a result, the system requires an external current supply to maintain this new electrochemical condition. The applied current (i_{applied}) needed to sustain this shift is calculated as the net current difference between the anodic and cathodic currents:

$$i_{\text{applied}} = |i_a| - |i_c|$$

This applied current corresponds to the control current in an anodic protection system. Once the metal enters the passive region, the anodic current drops sharply, and only a small current is needed to maintain the passive state. This makes anodic protection highly efficient for passivating metals in aggressive environments, as long as the metal's behavior permits a stable passive zone.

When the anodic polarization is increased to a value of ΔE_1, the system enters the unstable zone, a transitional region between the active and passive states. At this point, the anodic current (i_a) reaches the critical current density (i_{crit}).

This is the maximum current the metal experiences before passivation occurs, and it reflects a condition of high dissolution rate without the benefit of surface film formation. As a result:

- The applied current required to maintain this state is equal to the critical current:

$$i_{\text{applied}} = i_{\text{crit}}$$

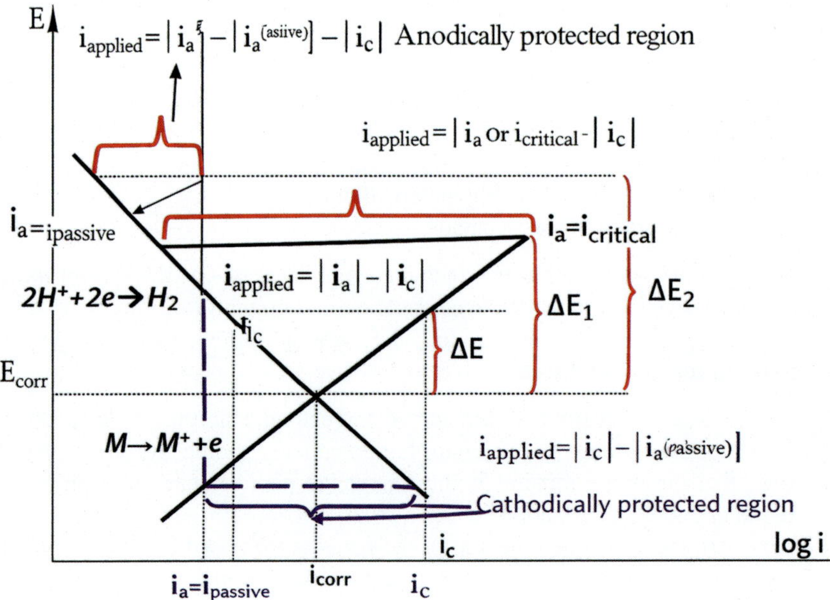

Fig. 9.6 Anodic protection

- This condition demands a substantial power input, since the metal is undergoing active corrosion, and no passive film has yet formed to reduce current demand.
- Therefore, operating in this unstable zone is inefficient and undesirable, as the energy requirements are high, and the surface is not yet protected.

Only once the potential is increased beyond ΔE_1, into the passive region, does the anodic current sharply drop, allowing the system to operate at a low, stable passivation current—the optimal condition for anodic protection.

$$i_{\text{applied}} = \left|i_a \text{ OR } i_{\text{critical}}\right| - \left|i_c\right|$$

When the anodic polarization increases further to ΔE_2, the metal enters the passivation zone. In this region, the anodic current (i_a) sharply decreases and stabilizes at a very low value, known as the passivation current (i_{passive}).

At this point:

- The applied current required to maintain the passivated state is:

$$i_{\text{applied}} = i_{\text{passive}}$$

- Although the anodic polarization (ΔE_2) is higher than ΔE_1, the corresponding anodic current is much lower.
- Compared to the critical current at ΔE_1, the passivation current at ΔE_2 is significantly smaller, resulting in lower energy demand.

This condition reflects the ideal operating region for anodic protection: the metal is fully protected by a stable oxide film, and only a minimal external current is needed to sustain the passive state, making the system highly efficient both electrically and chemically.

In the case of active-passive metals, it is generally more efficient and economical to use anodic protection rather than cathodic protection to achieve the same level of corrosion control. This recommendation is based on the electrochemical behavior of such metals and the power requirements of each method.

As illustrated in Fig. 9.6, the cathodic current required to shift the potential of the metal into the passive region (through cathodic protection) is substantially greater than the passivation current (i_{passive}) required in anodic protection. This large difference in current translates to a significant increase in energy consumption when cathodic protection is applied.

In contrast:

- Anodic protection requires only a small, controlled current to maintain the metal in its passive state, once the potential is elevated above the critical passivation potential.
- The anodic current used in the passive region is very low (i_{passive}), making the system highly efficient and cost-effective over time.

- By comparison, cathodic protection would demand a much larger current (often orders of magnitude greater) to polarize the metal into a protective state, especially in high-resistance environments.

For active-passive metals, such as stainless steel, titanium, and chromium-containing alloys, anodic protection is preferred because it delivers effective corrosion control with resulting in significantly lower current and energy demand, which makes anodic protection a more economical and technically appropriate solution, especially for metals capable of passivation.

Advantages of Anodic Protection
Anodic protection offers several advantages, particularly for metals that exhibit passivation behavior in aggressive environments:

1. Effective in Highly Corrosive Environments:
   Suitable for strong acid media (e.g., sulfuric acid) where passive film formation is possible.
2. Corrosion Rate Monitoring:
   The corrosion rate is directly proportional to the applied current, allowing for real-time monitoring and control.
3. Low Current Requirement:
   Once passivation is achieved, only a very small current is needed to maintain protection, making the system energy efficient.
4. Precise Potential Control:
   Allows for accurate regulation of electrode potential, ensuring the metal remains in the passive zone.
5. Uniform Current Distribution:
   Provides excellent current distribution across the protected surface, reducing the risk of localized corrosion.

Disadvantages of Anodic Protection
Despite its benefits, anodic protection has limitations that restrict its applicability:

1. High Initial Cost:
   Requires complex control systems and precise instrumentation, leading to higher capital and maintenance costs.
2. Ineffective Above the Waterline:
   Protection is only possible where the metal is immersed in a conducting electrolyte; surfaces exposed to air are not protected.
3. Unsuitable for Non-Conducting Electrolytes:
   It cannot be applied in non-conductive environments, where no ionic current can flow.
4. Incompatibility with Galvanic Coupling:
   Galvanically coupled materials (metals with different electrochemical potentials) can disrupt the potential control, making anodic protection ineffective in such systems.

Applications of Anodic Protection

Anodic protection is primarily applied in industries where metals operate in highly corrosive environments and exhibit a clear passivation range. Typical applications include:

- **Paper Industry:**

Used to protect equipment exposed to sulfuric acid and other aggressive chemicals in the pulp bleaching and chemical recovery processes.

- **Phosphoric Acid Plants:**

Applied to storage tanks and process vessels made of carbon steel, where phosphoric acid presents a severe corrosion risk.

- **Ammonia Fertilizer Plants:**

Employed in systems utilizing carbon steel components exposed to highly acidic or corrosive environments, such as urea or nitric acid handling lines.

9.3 Cathodic Protection Design

Cathodic Protection Design: Key Steps

Designing an effective Cathodic Protection (CP) system involves several critical steps to ensure optimal corrosion control, structural longevity, and cost-efficiency. The process typically includes the following phases:

1. **Evaluating the Structure**

   The first step involves a comprehensive assessment of the structure to be protected. Important factors include:

   - Structure Geometry:

   Complex shapes may influence current distribution.

   - Bare Surface Area vs. Coated Surface Area:

   Determines current requirements and helps estimate corrosion risk.

   - Coating Type and Quality:

   A high-quality coating reduces current demand; coating holidays must be considered.

   - Electrical Continuity:

   The structure must be electrically continuous, i.e., there must be no insulating joints, so that the protection current can reach all exposed surfaces.

   - Material Composition (Metallurgy):

   Different alloys and metals have different corrosion potentials and current demands.

2. **Evaluating the Environment**

   Environmental conditions significantly affect the efficiency and performance of CP systems. Key considerations include:

- Electrolyte Resistivity/Conductivity:

Low-resistivity environments (e.g., seawater) are favorable for CP; high-resistivity soils may require higher driving voltage.

- Presence of Nearby Structures:

Other buried or submerged metallic structures may interact electrically with the CP system (risk of stray current).

- Environmental Variability:

Consider fluctuations in:

Fluid velocity (e.g., flowing seawater).
Oxygen concentration.
Sulfide-Reducing Bacteria (SRB) activity.
Temperature: Elevated temperatures (e.g., >40 °C) increase anode consumption rates; zinc anodes are typically unsuitable above 45 °C.
pH levels.

3. **Evaluating the Site**

Site-specific physical constraints and layout considerations impact anode placement and CP system design. Key factors include:

- Physical Space for Anode Installation:

Whether sufficient area exists for impressed current ground beds or distributed sacrificial anodes.

- Current Distribution Challenges:

Assess how far anodes can be located from the structure while still ensuring uniform protection.

- Power Source Availability:

Especially critical for Impressed Current Cathodic Protection (ICCP) systems.

4. **Evaluating the Current Required**

Estimating the current demand is essential for selecting anode size, number, and power requirements. This can be done by:

- Theoretical Calculation:

Based on the total surface area to be protected and the required current density, considering environmental and coating conditions.

- Field Testing (Current Demand Testing):

A temporary ground bed and measurement system can be used to evaluate the actual current required on an existing structure under site-specific conditions (Table 9.2).

Example (Sample Current Calculation)

Design cathodic protection for the bottom of an above-ground storage tank (AST).

Given:

- **Tank diameter** = 200 ft.
- **Current density (Cd)** = 1 mA/ft^2.

Table 9.2 illustrates the amount of current needed for cathodic protection of steel in different environments

Current density (mA/m^2)	Environment
1-2	Stagnant seawater, coated
20-30	Seawater, uncoated
2-5	Seawater (1 m/s), coated
50-150	Seawater (1 m/s), uncoated
5-7	Seawater (2 m/s), coated
150-300	Seawater (2 m/s), uncoated
1-2	Steel in soil (0.5–5 Ω·m)
1-0.5	Steel in soil (5–15 Ω·m)

Calculation

$$\text{Surface area}\left(\text{SA}\right) = \pi \mathbf{r}^2 = 3.14 \times \left(100\right)^2 = 31{,}400\,\text{ft}^2.$$

$$\text{Current required} = \text{SA}^* \, \text{C}_\text{d} = 31.4\,\text{A}.$$

Current Density Requirements

The required current density in cathodic protection design is not a fixed value; it varies based on multiple factors:

- Electrolyte Characteristics:

The more corrosive or conductive the environment, the higher the current density required to achieve protection. For example, seawater requires less driving voltage than high-resistivity soils.

- Material and Structural Type:

Different materials require different levels of current for protection:
  - Buried steel structures: approximately 1.3 mA/ft^2.
  - Buried copper structures: can require up to 20 mA/ft^2.

- Corrections for Coating and Temperature:
  - Poor or damaged coatings significantly increase current demand.
  - Higher temperatures (above 40 °C) accelerate corrosion and increase current density requirements.

1. **Anode Selection**

   Once the required current is determined, the next step is to select appropriate anodes, which involves:

   - Understanding the Design Life:

   Anodes must supply sufficient current over the entire design period, often 20–30 years for infrastructure applications.

- Current Distribution Considerations:

Ensure uniform distribution of protective current across the surface area.

- Electrolyte Compatibility:

The anode material must function effectively in the site's electrolyte conditions (e.g., soil, seawater, concrete).

- Circuit Resistance and Geometry:

The system's total resistance affects the required driving voltage and should be minimized.

- Physical Constraints:

Installation conditions may limit anode size, spacing, or depth, which should be factored into the design.

Anode selection includes both the anode material (e.g., zinc, magnesium, aluminum, or MMO) and the anode configuration/geometry (e.g., rods, ribbon, plates, mesh, deep ground beds).

2. **Anode System Resistance**

System resistance is a key factor in determining both the efficiency and power requirements of the CP system.

- The resistance between the anode and the surrounding electrolyte impacts the current output and voltage requirements.
- As a general rule:
  - Longer anodes tend to have lower resistance.
  - Higher-resistivity electrolytes (e.g., dry soils) increase system resistance.

Calculating anode system resistance helps optimize the number, type, and placement of anodes to achieve protection at minimum voltage and energy cost.

System resistance (Example)

Anode System Resistance: Dwight's Equation.

To design an efficient cathodic protection system, it is critical to calculate the anode system resistance (R_A), the resistance between the anode surface and the surrounding electrolyte. This resistance directly influences the required driving voltage and system performance.

One of the most widely used formulas for this purpose is Dwight's Equation, which provides an estimate of resistance based on anode geometry and soil resistivity.

$$R_A = \frac{\rho}{2\pi L}\left(\ln\frac{4L}{r} - 1\right)$$

Dwight's Equation for a Cylindrical Rod Anode:

For a cylindrical rod anode embedded in soil, with the following parameters:

- ρ = Soil resistivity (Ω·cm),
- L = Length of the anode (cm).
- r = Radius of the anode (cm).

Interpretation:

- Lower resistance (R_A) leads to higher current output and better system efficiency.
- Longer anodes (↑L) and larger diameters (↑r) reduce resistance, improving current distribution.
- Higher soil resistivity (↑ρ) increases resistance and may require more driving voltage or additional anodes.

For example, using Dwight's equation for a 4″ × 80″ graphite anode in 4000 ohm-cm soil

$$R_A = \frac{4000\,\text{ohm.cm}}{2\pi 80\,\text{in} \times 2.54\,\text{cm / in}} \left(\ln\left(\frac{4 \times 80}{2} \right) - 1 \right)$$

$$R_A = 13\,\text{ohms}$$

Total System Resistance

In a cathodic protection system, the total circuit resistance determines the amount of voltage needed to deliver the required protective current. This total resistance includes more than just the anode-to-electrolyte resistance.

The total system resistance (**R_{total}**) is given by:

$$R_{\mathbf{total}} = R_{\mathbf{Anode}} + R_{\mathbf{Cabling}+} R_{\mathbf{Structure}}$$

Where:

- $R_{\mathbf{Anode}}$ is the resistance between the anode and the surrounding electrolyte (often calculated using Dwight's Equation).
- $R_{\mathbf{Cabling}}$ is the resistance of the connecting wires between the power source and the anode/structure.
- $R_{\mathbf{Structure}}$ is the resistance of the protected structure (usually negligible due to its large cross-sectional area and good conductivity).

In most cases, $\mathbf{R_{Anode}}$ dominates the total resistance, while $\mathbf{R_{Cabling}}$ and $\mathbf{R_{Structure}}$ are relatively small and often considered negligible in preliminary design calculations.

Current Output and Voltage Requirement

The **required voltage** to achieve a desired current output in a cathodic protection system is governed by **Ohm's Law**:

Where:

- **V** = Voltage required (volts).
- **I** = Desired current output (amperes).
- **R** = Total circuit resistance (ohms).

$$V = IR$$

Example: Sizing the Rectifier

Anode: 4″ × 80″ graphite
Soil resistivity: 4000 Ω·cm
Calculated anode resistance (RA):13 Ω
Assumed cable and structure resistance: 1 Ω

Total system resistance:

$$\mathbf{R}_{\text{total}} = R_A + R_{\text{cabling}} + R_{\text{structure}} = 13 + 1 = 14\,\Omega$$

For the 4″ × 80″ graphite anode in 4000 ohm-cm soil with a R_a of 13 ohms and a desired output of 3 A, the rectifier would need to be sized for a minimum of:

$$V = 3 \times 14 = 42\,\text{volts.}$$

Assumes = 1 ohm for cable and structure resistance.

To achieve a **3-amp output** from the graphite anode in this system, the **rectifier must be capable of supplying at least 42 volts**. This ensures sufficient driving voltage to overcome total system resistance and deliver the protective current.

1. **Design life estimation**

   The life expectancy of an anode is calculated using its consumption rate and utilization efficiency. This determines how long the anode will continue to provide protective current before it is fully consumed.

   The general formula for **anode life (years)** is:

$$Design\ life = \frac{Weight \times Utilization\ factor * 1000}{Output \times Consumption\ rate * 8760}$$

Where:

- **W** = Anode weight (kg).
- **U** = Utilization factor (typically 0.85–0.95 depending on material and system design).
- **I** = Average current output (A).

- **8760** = Number of hours in a year
- **1000** = Conversion factor (kg to grams) or to align units depending on the current format.

For our 4″ × 80″ anode operating at 3 A with a weight of 72 lb, a consumption rate of 2 lb/A-year and an 80% utilization factor that results in a design life of:

Given:

- Anode weight (W): 72 lb.
- Current output (I): 3 A.
- Consumption rate (CR): 2 lb/A·year.
- Utilization factor (U): 80% = 0.80.

Calculate Effective Anode Weight

$$W_{\mathbf{effective}} = 72\ \text{lb} \times 0.80 = 57.6\ \text{lb W}$$

$$\mathit{Design\ life} = \frac{57.6}{2\frac{\text{lb}}{\text{A year}} \times 3\text{A}} = 9.6\ \mathit{years}$$

Wrapping Up: Cathodic Protection Design Basics

Cathodic Protection (CP) system design is both a quantitative and qualitative engineering process that requires a balance of calculation and experience.

Quantitative Aspects of CP Design

A properly trained CP designer can quantitatively calculate critical design parameters, including:

- Total surface area to be protected.
- Current requirements, either by:
  - Field testing, or
  - Applying appropriate current density factors based on the environment and material.
- Anode and system circuit resistance (e.g., using Dwight's Equation).
- Anticipated anode life, based on consumption rate and utilization factor.

Qualitative Aspects of CP Design

While calculations provide a solid foundation, current distribution is often complex and influenced by factors that defy precise analytical modeling. Therefore, successful design also depends on:

- Sound engineering judgment.
- Field experience.
- Understanding of soil/environmental variability.
- Proper anode placement and spacing.

Anode Distribution Configurations

CP systems use various anode configurations, selected based on structural geometry, environmental conditions, and accessibility. The three primary types include:

1. Remote Anodes.
   - Located at a distance from the structure; current reaches the structure through the electrolyte.
2. Discrete Anodes.
   - Placed at individual, localized points near the structure; often used in small or irregularly shaped installations.
3. Linear Anodes.
   - Installed along the length of a structure (e.g., pipelines); provides uniform current distribution over long distances.

Ground Bed Classifications

The term "ground bed" refers to anode installations placed in the ground. These are classified into two main types:

- **Shallow Ground Beds:**
  - Anodes installed close to the surface, typically used where space is limited or shallow protection is adequate.
- Deep Well Ground Beds:
  - Anodes installed in drilled boreholes typically ranging from 100 ft (33 m) to 500 ft (160 m) in depth.
  - Often used in high-resistivity environments or where space constraints exist.
  - Output ranges typically from 8 to 75 A, depending on system needs.

Chapter 10
Forms of Corrosion

10.1 Uniform Corrosion

Corrosion can manifest in various forms, each with distinct visual characteristics, mechanisms, and engineering implications. These forms are typically classified based on the appearance of the affected metal surface and the underlying electrochemical behavior.

There are generally 13 major forms of corrosion, each requiring different diagnostic and mitigation approaches in design and maintenance.

Uniform corrosion is characterized by the even distribution of corrosion across the entire exposed surface of a metal. In this form, the corrosion product (typically metal oxides) forms a consistent and uniform layer, as illustrated in Fig. 10.1.

Because the rate of material loss is uniform, the remaining life of the structure or component can be predicted with relative ease, making this one of the least dangerous and most manageable forms of corrosion in engineering practice.

Fig. 10.1 Uniform corrosion

M. M. Ghatus, *Fundamentals of Corrosion Science and Engineering*, https://doi.org/10.1007/978-3-032-13138-6_10

Common Examples

- Rusting of steel in atmospheric environments.
- Tarnishing of silverware due to uniform surface oxidation.
- Corrosion of car bodies and steel frameworks.
- Oxidation of structural steel exposed to air.

In most cases, such as rusting of steel in the air, both anodic and cathodic reactions occur on the same surface, resulting in a steady and predictable rate of material degradation.

$$Fe \rightarrow Fe^{2+} + 2e \quad \text{Anodic Reaction}$$
$$O_2 + 2H_2O + 4e \rightarrow 4OH^- \quad \text{Cathodic reaction}$$
$$4Fe(OH)_2 + O_2 + 2H_2O \rightarrow 4Fe(OH)_3$$
$$4Fe(OH)_3 + O_2 \rightarrow 2Fe_2O_3, H_2O + 2H_2O$$

Factors Contributing to Uniform Corrosion

Several environmental and chemical conditions influence the rate and severity of uniform corrosion. The most common factors include:

- Humidity:
  Moisture in the air facilitates the formation of thin electrolyte films on metal surfaces, promoting continuous corrosion reactions.
- Presence of Pollutants (H_2S, SO_2):
  Gaseous pollutants such as hydrogen sulfide (H_2S) and sulfur dioxide (SO_2) accelerate corrosion, particularly of metals like silver, leading to tarnishing.
- Temperature:
  Corrosion rates typically increase with rising temperature due to accelerated electrochemical and diffusion processes.
- Salt Contamination (e.g., NaCl):
  Chloride ions from salts, especially sodium chloride (NaCl), disrupt passive films and act as aggressive corrosion promoters in marine and coastal environments.
- Dew Formation:
  Condensation on metal surfaces, especially in fluctuating temperatures, provides a localized electrolyte layer, intensifying corrosion activity.

Protection Against Uniform Corrosion

Uniform corrosion, while predictable, still requires effective preventive measures to ensure structural integrity and prolonged service life. The most common and practical protection methods include:

- Allowance for Extra Material (Corrosion Allowance):

Structural components can be designed with increased thickness, accounting for the expected material loss over time due to corrosion. This approach is commonly used in pipelines, tanks, and offshore structures.

- Cathodic Protection (CP):

CP is a highly effective technique for controlling uniform corrosion, particularly in buried or submerged metallic structures. Two common types are:
  - Impressed Current Cathodic Protection (ICCP).
  - Sacrificial Anode Cathodic Protection (SACP).

- Protective Coatings and Painting:

Applying paint or corrosion-resistant coatings isolates the metal from its corrosive environment, thereby significantly reducing corrosion rates. Periodic inspection and maintenance are essential for long-term effectiveness.

10.2 Galvanic Corrosion

Galvanic corrosion occurs when two dissimilar metals are electrically connected in the presence of a common electrolyte. Based on the galvanic series for a specific environment (e.g., seawater), one metal is more noble (cathodic) and the other more active (anodic). The anodic metal corrodes preferentially to protect the cathodic metal.

For example, when aluminum (Al) and copper (Cu) are coupled in an electrolyte (as illustrated in Fig. 10.2):

- Aluminum, being more active, becomes the anode.
- Copper, being nobler, becomes the cathode.
- The result is accelerated corrosion of aluminum, while copper is protected.

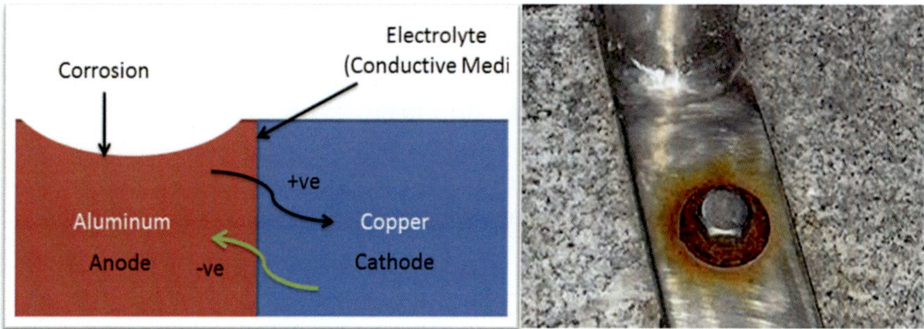

Fig. 10.2 Galvanic corrosion

Practical Examples of Galvanic Corrosion

- A Monel (nickel-copper alloy) yacht hull fastened with steel rivets: the rivets corrode.
- A copper plate joined with steel rivets: again, the rivets corrode.
- Steel bolts in contact with brass or bronze components in marine or plumbing systems.

In both Monel and copper examples, steel serves as the anode and corrodes faster due to its more active position in the galvanic series.

Characteristics of Galvanic Series

The galvanic series is an experimentally derived ranking of metals and alloys according to their corrosion potentials in a specific electrolyte, typically seawater. It serves as a practical guide for predicting galvanic compatibility between dissimilar metals.

Key characteristics include:

1. Correlation with the Electromotive Force (EMF) Series:
   The position of a metal in the galvanic series generally corresponds to its position in the standard electromotive force series, though real-world conditions (e.g., surface condition, electrolyte type) influence actual performance.
2. Effect of Passivity:
   The formation of a passive film can significantly alter a metal's position in the series. For instance, stainless steel may act more nobly in a passivated state but more actively if passivity is disrupted.
3. Bracketed Positions:
   Metals listed within brackets in the series have very similar corrosion potentials, indicating they are galvanically compatible. When such metals are coupled, galvanic corrosion is unlikely to occur.

Factors That Affect Galvanic Corrosion

The severity of galvanic corrosion is significantly affected by environmental conditions, particularly those related to the electrolyte in which the metals are immersed.

Salt Content (Electrolyte Conductivity)

- Increased salt concentration (e.g., NaCl in seawater) enhances the electrolyte's conductivity, allowing current to flow more easily between the anode and cathode.
- This results in an accelerated corrosion rate at the anodic area, as more electrons are released and more metal is consumed.
- In contrast, low-conductivity or non-conductive electrolytes introduce a higher IR (voltage) drop across the system.
  - The anodic current is reduced, which slows down the corrosion rate.
  - This effect is illustrated in Fig. 10.3, where poor conductivity leads to a diminished current flow and hence lower galvanic corrosion intensity.

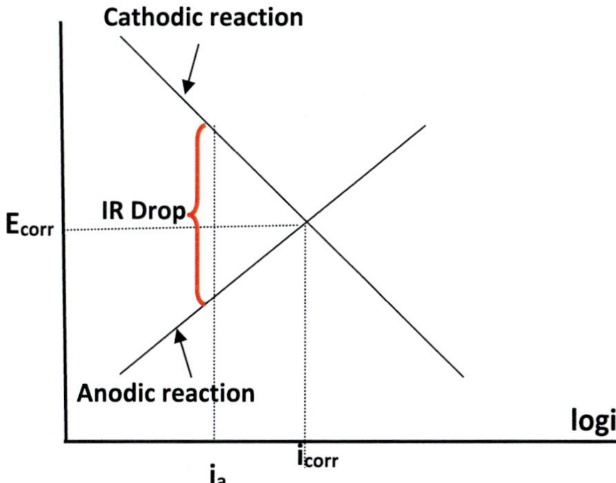

Fig. 10.3 Effect of salt on corrosion rate

Temperature

Temperature plays a critical role in the rate and severity of galvanic corrosion, especially for galvanized iron (steel coated with zinc).

- At elevated temperatures, corrosion reactions become thermodynamically and kinetically more favorable, accelerating both anodic metal dissolution and cathodic reduction reactions.
- For galvanized iron, if the zinc coating is scratched or damaged in the presence of water, zinc begins to act as the anode, while the exposed steel acts as the cathode.
- The cathodic reaction in water generates hydroxide ions (OH^-):

$$O_2 + 2H_2O + 4e^- \rightarrow 4OH^-$$

- Zinc, being more active, undergoes oxidation, protecting the steel:

$$Zn \rightarrow Zn^{2+} + 2e$$

As temperature increases:

- The rate of zinc dissolution increases.
- Hydroxide production accelerates.
- The zinc coating depletes faster, reducing the protection time.

Higher temperatures significantly reduce the service life of galvanized coatings by intensifying galvanic activity, especially at coating defects.

At temperatures up to 180 °F (≈82 °C), zinc ions (Zn^{2+}) released during the corrosion of galvanized iron react with hydroxide ions (OH^-) generated from the cathodic reaction to form zinc hydroxide:

$$Zn^{2+} + 2OH^- \rightarrow Zn(OH)_2$$

In this temperature range:

- Zinc (Zn) acts as the anode, protecting the underlying iron (Fe), which functions as the cathode.
- Zinc hydroxide [$Zn(OH)_2$] deposits as a protective film on the surface, contributing to corrosion resistance.

However, when the temperature exceeds 180 °F, a transformation occurs:

- Instead of forming zinc hydroxide, zinc oxide (ZnO) forms as a powdery corrosion product.
- Under these high-temperature conditions, the corrosion mechanism may reverse:
  - Iron (Fe) becomes the anode.
  - Zinc oxide (ZnO) behaves as the cathode.

This reversal in anodic and cathodic behavior is known as reversible polarity, a phenomenon that significantly reduces the protective effectiveness of the zinc coating and accelerates corrosion of the steel substrate at elevated temperatures.

Area Ratio (Cathodic Area/Anodic Area)

In galvanic corrosion, the area ratio between the cathode and anode plays a critical role in determining the corrosion rate of the anodic material.

When two dissimilar metals are electrically connected in an electrolyte, the anode corrodes while the cathode is protected. The rate at which the anode corrodes is inversely proportional to its surface area and directly influenced by the cathodic demand.

Key Principles

- If the cathodic area is large relative to the anodic area, the anode must supply more electrons to support the cathodic reaction. This results in:
  - A higher anodic current density.
  - A faster corrosion rate at the anode (unfavorable condition).
- Conversely, if the anodic area is large compared to the cathodic area, the corrosion rate per unit area of the anode is reduced (favorable condition).

Example: Zinc Coupled with Platinum in a Hydrogen Environment

(*Illustrated in* Fig. 10.4)

In this case, zinc (Zn) acts as the anode, and platinum (Pt) acts as the cathode. The environment contains hydrogen, which participates in cathodic hydrogen evolution reactions.

- Two hydrogen cathodic reactions occur simultaneously:
  - One on the zinc surface.
  - One on the platinum surface.

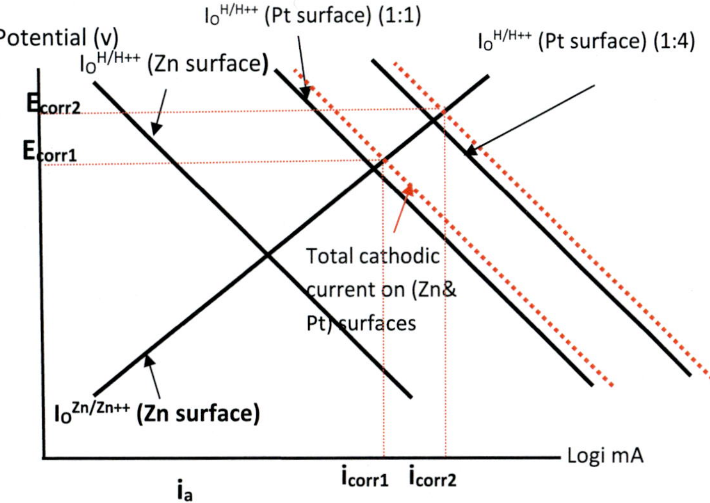

Fig. 10.4 Effect of area ratio on corrosion rate of coupled Zn-Pt

- The dashed line in Fig. 10.4 represents the summation of both cathodic reactions.

Case 1: Area Ratio = 1:1 (Zn:Pt)
- The cathodic reaction on platinum is moderate.
- The corrosion current is I_{corr1}.

Case 2: Area Ratio = 1:4 (Zn:Pt)
- The cathodic area (Pt) is increased.
- The hydrogen reduction current on Pt shifts to the right (increased).
- This increases the total cathodic current.
- The new corrosion current becomes $I_{\mathrm{corr2}} > I_{\mathrm{corr1}}$, indicating a higher corrosion rate on the zinc anode.

To minimize galvanic corrosion, it is essential to design systems where the anodic area is large relative to the cathodic area, especially when protective coatings or dissimilar metals are involved.

Therefore, surface area plays a direct and critical role in determining the corrosion rate in galvanic systems. As previously discussed, when the cathodic area increases relative to the anodic area, the demand for electrons rises, forcing the anode to supply more current. This results in a higher anodic current density and, consequently, an accelerated corrosion rate at the anodic site.

Designing with an optimized area ratio—where the anodic area is equal to or greater than the cathodic area—is essential to mitigate galvanic corrosion effectively.

Distance Effect
The distance between dissimilar metals in a galvanic couple plays a significant role in the distribution of corrosion activity, particularly on the anodic surface.

When two dissimilar metals are electrically connected in the presence of an electrolyte, the corrosion rate is not uniformly distributed across the anodic metal. Instead, it is highest near the contact interface and decreases with increasing distance from that interface.

- At the contact surface (the junction or connection line) between the anode and the cathode, electron exchange and ionic transport are most efficient, resulting in a high corrosion rate.
- As the distance from the contact surface increases, the resistance to ionic and electronic flow increases, causing the corrosion rate to decline.

This phenomenon is illustrated in Fig. 10.5, where:

- Metal 2, being more active, functions as the anode.
- Metal 1, being more noble, acts as the cathode.
- The corrosion intensity on Metal 2 is greatest at the interface and diminishes with distance from the contact line.

In galvanic couples, corrosion is most severe near the physical junction between dissimilar metals. Designers must account for this localized attack when planning material connections, especially in harsh environments.

Protection Against Galvanic Corrosion

To minimize or prevent galvanic corrosion, especially in environments where dissimilar metals must be used, several practical and engineering strategies can be applied:

1. Ideal Selection of Metals and Alloys
    - Select metals and alloys that are close in the galvanic series to reduce potential difference and limit galvanic current.
    - Prefer galvanically compatible materials when direct contact is unavoidable.
2. Avoid Unfavorable Area Ratios
    - Avoid configurations where a small anodic area is coupled with a large cathodic area, as this significantly increases the corrosion rate at the anode.
    - Designs should maximize the anodic area or minimize the cathodic area wherever possible.

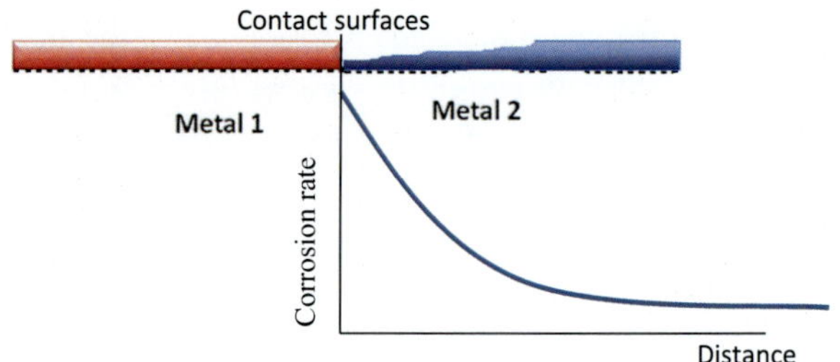

Fig. 10.5 Corrosion rate and distance from surface contact

3. Use of Coatings

   - Apply protective coatings, especially on the cathodic metal, to reduce its exposure to the electrolyte and limit cathodic current demand.
   - Caution: Coating only the anode may be ineffective or even harmful if coating defects become initiation sites for localized corrosion.

4. Corrosion Inhibitors

   - Add chemical inhibitors to the electrolyte to reduce overall corrosion activity and limit galvanic interactions.
   - Common inhibitors include chromates, phosphates, or organic film-forming compounds.

5. Cathodic Protection

   - Employ sacrificial anodes or impressed current cathodic protection (ICCP) systems to supply electrons and prevent galvanic attack on the protected structure.
   - Often used for buried pipelines, marine structures, and storage tanks.

Preventing galvanic corrosion involves a combination of material selection, geometric design, surface treatment, and, when necessary, electrochemical protection. Understanding the system's environment and electrical connectivity is key to effective mitigation.

Advantages of Galvanic Corrosion
Although galvanic corrosion is typically considered undesirable in engineering systems, it also has several practical and beneficial applications where the underlying electrochemical principles are intentionally exploited:

1. Electrical Energy Generation (Galvanic Cells/Batteries)

   - Galvanic corrosion forms the basis of electrochemical cells, where controlled oxidation and reduction reactions are used to generate electrical power.
   - Examples include zinc-carbon, lead-acid, and lithium-ion batteries.

2. Cathodic Protection Systems

   - Galvanic principles are used to protect critical metal structures from corrosion:

     – Sacrificial Anode Systems (SACP): A more active metal corrodes to protect the structure.
     – Impressed Current Cathodic Protection (ICCP): An external power source applies current to prevent corrosion.

3. Cleaning of Silverware

   - Galvanic action is used in household applications to remove tarnish from silverware.

- In this method, silver items are placed in contact with aluminum foil in a baking soda solution, where aluminum oxidizes and silver is reduced, effectively removing silver sulfide tarnish.

$$
\begin{aligned}
&3Ag_2S + 2Al \rightarrow 6Ag + Al_2S_3 \\
&Al \rightarrow Al^{3+} + 3e \\
&Ag^+ + e \rightarrow Ag \\
&Al_2S_3 + 6H_2O \rightarrow 2Al(OH)_3 + 3H_2S \uparrow
\end{aligned}
$$

While galvanic corrosion is often a problem to be mitigated, its controlled use plays a critical role in electrical energy systems, corrosion control technologies, and even domestic cleaning applications.

10.3 Crevice Corrosion

Crevice corrosion is a highly localized form of corrosion that occurs in shielded areas or narrow gaps where the access of the electrolyte is restricted and fluid flow is stagnant. This creates an environment that promotes differential aeration and localized chemical changes, leading to intense attack in the crevice zone.

Classical Definition

Crevice corrosion is defined as the intensification of localized corrosion, typically occurring within crevices and other confined geometries on metal surfaces that are exposed to corrosive environments.

Common Crevice Locations

- Between metal and non-metal surfaces (e.g., under gaskets, washers, or seals).
- Between two metals in bolted or riveted assemblies.
- In metal lap joints, seams, or folds.

The environment within a crevice becomes **stagnant**, which:

- Restricts oxygen diffusion.
- Promotes acidification.
- Leads to metal ion buildup.
- Initiates localized anodic dissolution.

When two different metals are involved, the galvanic effect can further intensify corrosion. Even when only one metal is present, differential concentration cells form due to the difference in oxygen availability between the crevice and the surrounding surface. This leads to a concentration cell effect, accelerating attack within the crevice.

As shown in Fig. 10.6, crevice corrosion often forms beneath gaskets or at contact points within structural components where stagnant electrolytes are trapped.

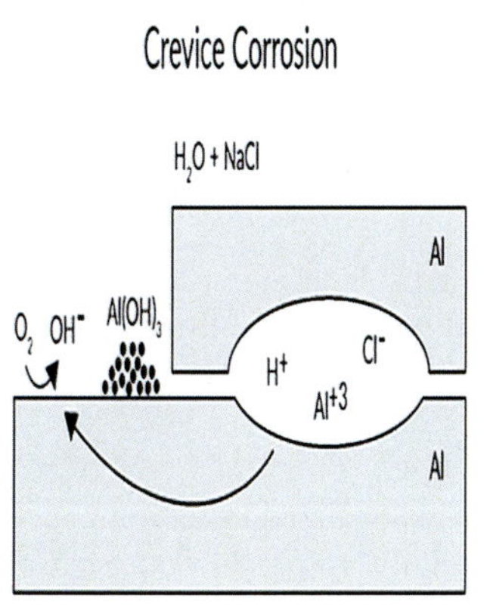

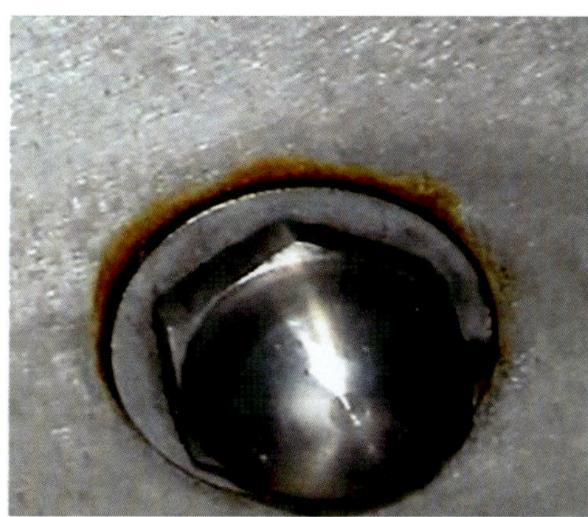

Fig. 10.6 Crevice corrosion

Causes and Governing Factors of Crevice Corrosion

Crevice corrosion arises from a unique set of physical and electrochemical conditions that develop within confined, stagnant environments, typically in narrow gaps or shielded areas. Both design geometry and environmental conditions play crucial roles in its initiation and propagation.

Common Causes of Crevice Corrosion

- Narrow gaps at metal-to-metal or metal-to-nonmetal contact points (e.g., under gaskets or washers).
- Cracked components, which create local crevices.
- Biofouling deposits on submerged surfaces, especially under seawater conditions.
- Accumulated debris, such as dirt, mud, or inorganic deposits, which block flow and trap corrosive agents.

Factors Governing Crevice Corrosion

1. Bulk Solution Composition:
   - Differences in oxygen concentration (differential aeration) between inside and outside the crevice create concentration cells.
2. Mass Transport Processes:
   - Includes migration, diffusion, and convection.
   - Example: Chloride ions (Cl^-) in seawater may diffuse into the crevice, accelerating corrosion.
3. Chemical Changes Inside the Crevice:
   - Progressive acidification (lower pH).
   - Increase in chloride ion concentration.
   - Depletion of oxygen, leading to reduced cathodic reaction and enhanced anodic dissolution.
   - Stagnation ensures these imbalances are maintained.

4. Electrochemical Reactions:
   - The crevice environment becomes anodic, while the surrounding surface becomes cathodic, driving localized attack.
5. Alloy Composition:
   - Alloys that form stable passive films are more resistant.
   - Crevice corrosion is more severe in alloys susceptible to film breakdown in low-pH or high-Cl^- environments.
6. Passive Layer Quality:
   - A weakened or damaged passive film promotes localized initiation.
   - Stainless steels and passive alloys are especially vulnerable once passivity breaks down.
7. Crevice Geometry:
   - Crevices must be:
     - Wide enough to permit water or ion entry.
     - Narrow enough to restrict flow, maintaining a stagnant zone.
8. Surface Area Ratio (Crevice vs. Total Area):
   - A small crevice area relative to the exposed surface promotes current localization, accelerating the corrosion rate inside the crevice.

Crevice corrosion is governed by a combination of microenvironmental instability, electrochemical imbalance, and physical confinement. Preventing this form of corrosion requires careful material selection, design detailing, and maintenance practices to avoid crevice-forming configurations.

Mechanism of Crevice Corrosion

Crevice corrosion proceeds through a series of electrochemical and chemical transformations that are driven by oxygen depletion, pH reduction, and passive film breakdown within the stagnant microenvironment of the crevice.

Stage 1: Oxygen Depletion Inside the Crevice

Initially, **metal dissolution** occurs within the crevice, releasing metal ions and electrons. For example, with aluminum:

$$Al \rightarrow Al^{3+} + 3e^-$$

The electrons travel to the cathodic area (typically outside the crevice), where they are consumed in the oxygen reduction reaction:

$$O_2 + 2H_2O + 4e^- \rightarrow 4OH^-$$

As the reaction proceeds, oxygen inside the crevice becomes depleted due to the stagnant condition. Since oxygen cannot be replenished, this leads to the accumulation of metal ions and the breakdown of electrochemical balance inside the crevice.

Furthermore, due to an unfavorable area ratio (large cathode vs. small anode), the anodic current density increases, accelerating corrosion within the crevice.

Stage 2: pH Reduction Due to Hydrolysis

To maintain electroneutrality, metal ions inside the crevice react with water molecules, resulting in the generation of hydrogen ions (H^+):

$$2/3Al^{3+} + 2H_2O \rightarrow 2/3Al(OH)_3 \downarrow + 2H^+$$

Additionally, chloride ions (Cl^-) in seawater react with aluminum ions to form aluminum chloride, which further hydrolyzes:

$$Al^{3+} + 3Cl^- \rightarrow AlCl_3$$
$$AlCl_3 + H_2O \rightarrow Al(OH)_3 + 2(H^+, Cl^-)$$

These reactions significantly lower the pH inside the crevice, making it highly acidic, which promotes further metal dissolution and loss of passivity.

Stage 3: Passive Layer Breakdown

In acidic and chloride-rich environments, metals like aluminum lose their passive oxide layer, which normally protects them from corrosion. This passivity breakdown initiates an intense, localized attack.

Stage 4: Propagation of Crevice Corrosion

As acidity increases and the passive film continues to degrade, the corrosion rate escalates. The system may eventually develop perforations or leaks, as the anodic zone expands and becomes more reactive, leading to structural failure.

Protection Against Crevice Corrosion

To mitigate or prevent crevice corrosion, the following design and maintenance strategies are recommended:

1. Avoid Crevice Geometry:
   - Use butt joints instead of lap joints.
   - Minimize fasteners or overlaps that trap electrolytes.
2. Proper Design Practices:
   - Incorporate rounded corners instead of sharp ones, particularly in water storage tanks or submerged structures.
3. Routine Cleaning and Maintenance:
   - Remove accumulated precipitates, sludge, or biofouling from surfaces such as tank bottoms and pipe interiors to prevent crevice formation.

Crevice corrosion is a progressive and localized attack that can be catastrophic if left unchecked. A combination of smart design, material selection, and preventive maintenance is critical to minimize its risk.

10.4 Pitting Corrosion

Pitting corrosion is a severe and highly localized form of corrosion that results in the formation of small cavities or pits on a metal surface. Unlike uniform corrosion, which occurs over a large area, pitting corrosion is concentrated in discrete spots, often penetrating deep into the material while leaving the surrounding surface relatively unaffected.

Key Characteristics

- Commonly affects passive metals, such as aluminum, stainless steel, and nickel-based alloys.
- Often occurs under protective films or coatings that have been locally damaged.
- Frequently initiated by chloride ions in environments like seawater, humid air, or acidic solutions.

As illustrated in Fig. 10.7, pitting appears as crater-like defects that may perforate the material, especially in thin sections or sheets.

Challenges and Dangers

- Unmeasurable and unpredictable:

The overall weight loss is minimal and does not reflect the true severity of the attack.

- Silent progression:

Pitting often progresses without visible warning, making it catastrophic when failure suddenly occurs.

- Difficult to detect:

Surface may appear unaffected while deep pits form beneath.

Measurement Metric

Since weight loss is not a reliable indicator, the penetration depth is used to characterize pitting severity. Common units include:

- mpy—mils per year,
- mmpy—millimeters per year.

Pitting corrosion is dangerous because it is highly localized, unpredictable, and can lead to sudden structural failure. Early detection and prevention are critical in systems using passive alloys.

Factors Affecting Pitting Corrosion

Pitting corrosion is influenced by a variety of material-related, processing, and environmental factors. These factors can either initiate pit formation or accelerate pit propagation, especially in passive metals like stainless steel and aluminum alloys.

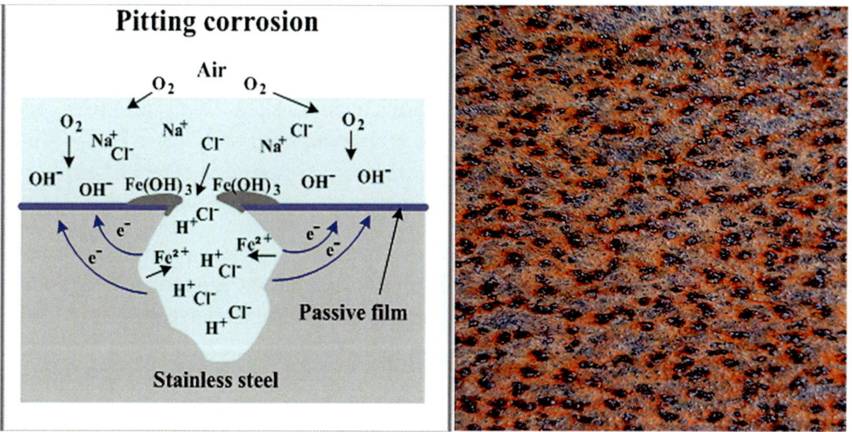

Fig. 10.7 Pitting corrosion

Surface Defects (e.g., Inclusions)

- Non-metallic inclusions or microstructural discontinuities act as initiation sites for pitting.
- These localized heterogeneities disrupt the passive film, making the surface more vulnerable to chloride attack.

Degree of Cold Work

- Cold working introduces residual stresses and dislocation networks into the metal structure, which reduce passivity stability.
- For example, austenitic stainless steel exposed to ferric chloride ($FeCl_3$) is particularly susceptible to pitting in cold-worked regions.
- Increased cold work enhances localized electrochemical activity, promoting pit nucleation and growth.

Surface Finish

- Surface finish plays a critical role in pitting resistance:
  - Smooth, highly polished surfaces with minimal scratches tend to offer better pitting resistance due to uniform passive film formation.
  - However, if the polished surface is too smooth or contains latent surface defects, it may promote localized breakdown under aggressive environments.

- Over time, even minute surface irregularities can evolve into deep pits if exposed to chlorides.

Sensitization Temperature (Grain Boundary Attack)

- Sensitization occurs when chromium carbide precipitates form at grain boundaries in stainless steel (e.g., Type 18-8) during slow cooling through the 400–600 °C range.

- This causes chromium depletion in the adjacent zones, creating a local anode surrounded by large cathodic regions, a condition favorable for pitting and intergranular corrosion.
- The affected areas along grain boundaries become sites for pit initiation, leading to possible structural failure along the grain boundary path.
- Solution: Use stabilized grades (e.g., 321 or 347 stainless steel) or apply rapid cooling (quenching) after solution treatment to prevent sensitization.

A combination of surface integrity, mechanical history, and thermal treatment significantly affects the pitting behavior of passive alloys. Understanding and controlling these factors is critical to designing corrosion-resistant systems.

Environmental and Material Factors Influencing Pitting Corrosion

- **Environmental Contamination**

Pitting corrosion is often associated with crevice corrosion, as both share similar mechanisms driven by localized stagnation and electrochemical imbalances. While pitting can occur independently, crevices almost always initiate pitting.
  - Contaminants like biofouling, dirt, or debris promote stagnation and trap corrosive agents such as chlorides.
  - Preventive Action: Regular cleaning of surfaces to remove bio-organic matter, sediments, and marine growth is essential in preventing pit initiation.

- **Alloy Composition**
  - Chromium (Cr) content is vital in resisting pitting. A homogeneous distribution of Cr in stainless steel forms a stable passive film.
  - Molybdenum (Mo) enhances pitting resistance, particularly in chloride-rich environments, by reinforcing the passive layer.

  For example, stainless steel 18-8 (18% Cr, 8% Ni) gains improved pitting resistance when Mo is added.
- **Flow Velocity**
  - Fluid motion discourages pitting by continuously removing aggressive ions from the surface and replenishing oxygen.
  - Stagnant conditions trap chloride ions and promote film breakdown.
  - Conclusion: Pitting likelihood is inversely proportional to fluid velocity. Avoiding stagnant zones is critical in design.

Conditions Favoring Pitting Corrosion

Pitting is triggered under specific electrochemical and environmental conditions:

- Breakdown of the Passive Film.
  - Caused by chemical inhomogeneity, mechanical damage, or differential aeration.

- Presence of Halide Ions.
  - Particularly Cl^-, Br^-, I^-, and F^-, which penetrate and destabilize passive layers.

- Stagnant Electrolytes.

– Enable localized ion buildup and oxygen depletion inside the pit.

Mechanism of Pitting Corrosion

The mechanism of pitting corrosion closely resembles that of crevice corrosion, involving three critical prerequisites and three distinct stages.

Preconditions

1. Dissolved Oxygen (to enable cathodic reaction).
2. Passivating Metal (e.g., Aluminum or Stainless Steel).
3. Electrolyte with Halides (e.g., seawater containing Cl^-).

Stages of Pitting Corrosion

1. Pit Initiation
   - Initiated at sites with defects, grain boundary depletion, or inclusions.
   - For instance, depleted grain boundary (DGB) zones in sensitized stainless steel (chromium-depleted regions) become anodic sites.
   - These local areas have higher surface energy, leading to accelerated dissolution.
2. Pit Propagation
   - Once the pit begins, the local environment becomes acidic and enriched with metal chlorides.
   - Anodic Reaction:

   (e.g., for iron):

$$Fe \rightarrow Fe^{2+} + 2e^-$$

   - Cathodic Reaction (outside the pit):

$$O_2 + 2H_2O + 4e^- \rightarrow 4OH^-$$

   Hydroxide ions react with metal ions to form:

$$Fe^{3+} + 3OH^- \rightarrow Fe(OH)_3 \downarrow$$

   Stagnancy within the pit causes oxygen depletion, forming a concentration cell
   - Based on the Nernst equation, the partial pressure of oxygen is greater outside the pit (cathodic area) than inside (anodic area), intensifying the corrosion rate.
3. Pit Termination
   - Eventually, the pit may either self-repassivate if conditions stabilize or continue growing until material perforation occurs.

Pitting corrosion is driven by a breakdown in the passive film and sustained by localized electrochemical conditions. It is highly influenced by environmental contamination, material composition, and stagnant electrolyte conditions. Early detection and preventive design are essential due to its stealthy and destructive nature.

$$E = E^\circ + \frac{RT}{4F} \ln \frac{P_{O2}\left(H_2O\right)^2}{\left(OH^-\right)^4}$$

Therefore, the electrochemical potential in the cathodic area is greater than in the anodic (pit) area, creating a potential gradient that sustains electron flow and accelerates anodic dissolution. As a result, an unfavorable area ratio develops a large cathode (surrounding surface) and a small anode (pit base), which drives a high localized current density and leads to a rapid increase in the corrosion rate within the pit.

The system attempts to achieve charge neutrality, but due to the presence of chloride ions (Cl^-), further chemical reactions occur:

1. Metal ions hydrolyze, generating hydrogen ions (H^+).
2. Chloride ions complex with metal ions, and then react with H^+ to form hydrochloric acid (HCl).

These reactions mirror those in crevice corrosion and cause the pH inside the pit to drop sharply, creating an acidic and aggressive environment. This low-pH environment further destabilizes the passive film and accelerates the metal dissolution process.

Ultimately, the pitting process propagates until the metal wall is perforated or structurally compromised, leading to unexpected and catastrophic failure.

Protection Against Pitting Corrosion

Due to its localized and unpredictable nature, pitting corrosion requires proactive strategies aimed at both preventing pit initiation and limiting pit propagation. Many preventive measures overlap with those used for crevice corrosion.

- **Adopt Crevice Corrosion Prevention Techniques**
  - Since pitting and crevice corrosion share similar mechanisms, all design strategies to avoid crevice formation, promote flow, and minimize deposits also help prevent pitting.
- **Stabilize the Passive Layer**
  - Use alloying elements that enhance the durability of the passive film:
    Molybdenum (Mo) additions to stainless steels (e.g., 316) significantly increase resistance to chloride-induced pitting.
    Nitrogen (N) and chromium (Cr) also improve passive film integrity.
  - Ensure the passive film remains uniform and free from mechanical or chemical disruption.

- **Avoid Electrolyte Stagnancy**
  - Design systems to maintain adequate fluid velocity, especially in piping and tank systems, to prevent localized buildup of chlorides.
  - Avoid dead zones or recessed geometries where fluid circulation is limited.
- **Select Appropriate Materials for the Environment**
  - Match materials to the chemical and thermal characteristics of the service environment.

    For example, use duplex stainless steels, super austenitic alloys, or titanium in chloride-rich or marine environments.
- Consider pitting resistance equivalent (PRE) numbers as a guide when selecting alloys for aggressive conditions.

Conclusion: The best protection against pitting corrosion lies in a combination of informed material selection, sound design, proper fluid dynamics, and passive layer reinforcement through alloying.

10.5 Intergranular Corrosion

Intergranular corrosion is a form of localized attack that occurs preferentially along grain boundaries in polycrystalline metals. While the grains themselves may remain unaffected, the grain boundary zones become highly active, making them susceptible to selective dissolution. This attack compromises structural integrity and can lead to grain detachment and premature failure, even when the bulk material appears corrosion-resistant.

As illustrated in Fig. 10.8, intergranular corrosion typically initiates at the grain boundary network due to metallurgical inhomogeneities introduced during processing or service.

Factors Contributing to Grain Boundary Activity

1. Enrichment of Reactive Elements:
   - For example, zinc enrichment at grain boundaries in brass promotes localized galvanic attack.
2. Impurity Segregation:
   - Iron impurities in aluminum may preferentially segregate at grain boundaries, making these regions anodic.
3. Depletion of Alloying Elements (Sensitization):
   - This is most commonly seen in austenitic stainless steels, where chromium depletion adjacent to grain boundaries leads to loss of passivity.

Fig. 10.8 Intergranular corrosion

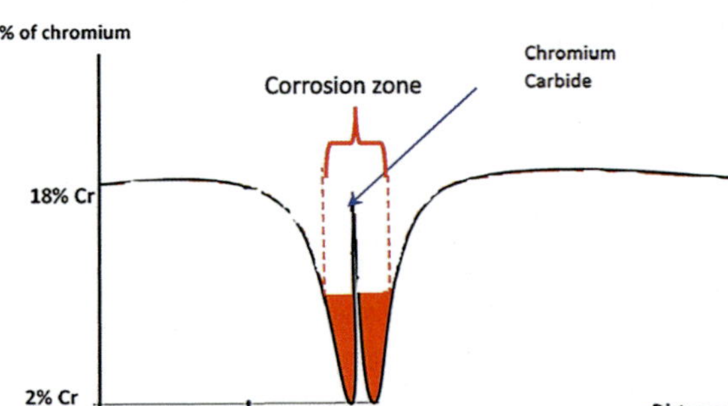

Fig. 10.9 Concentration of chromium and distance

Sensitization in Austenitic Stainless Steel (18.8)

- 18.8 stainless steel contains 18% chromium and 8% nickel in a solid solution.
- When subjected to slow cooling or long-term exposure in the 400–600 °C temperature range, chromium carbides ($Cr_{23}C_6$) precipitate along grain boundaries.
- The region adjacent to these carbides becomes depleted in chromium (<2%), falling below the threshold necessary to maintain a protective passive film.

As shown in Fig. 10.9, this results in a sharp drop in chromium content near the carbide, making those zones anodic, while the unaffected grains remain cathodic. This unfavorable area ratio (small anodic zone surrounded by a large cathodic matrix) accelerates localized corrosion along the grain boundary.

Key Conditions for Chromium Carbide Formation

- Carbon Content:
  - Chromium carbide forms when carbon content exceeds ~0.02%, based on the Fe-Cr-C phase diagram.

- Critical Temperature Range:
  - Sensitization occurs between 600 and 800 °C. This promotes carbide nucleation and growth, especially during slow cooling or long-term exposure.
- Carbon Content Example:
  - With 0.08% carbon, sensitization is more likely unless mitigated.

Prevention Through Alloying and Heat Treatment

- Rapid Cooling (e.g., quenching) from solution-annealing temperatures prevents chromium carbide formation, regardless of carbon content.
- Stabilizing Elements:
  - Niobium (Nb) and titanium (Ti) form niobium carbide (NbC) or titanium carbide (TiC), which preferentially consume excess carbon.
  - These carbides prevent chromium from reacting with carbon, thereby maintaining passivity and eliminating sensitization.

Intergranular corrosion is strongly influenced by heat treatment history, alloy composition, and grain boundary chemistry. Preventing sensitization through proper thermal practices and alloy stabilization is essential in ensuring the corrosion resistance of materials such as stainless steels.

Prevention and Control of Sensitization in Austenitic Stainless Steels

Sensitization is a condition where chromium carbide precipitates at grain boundaries, leading to chromium depletion and intergranular corrosion. It can be effectively prevented or controlled through the following measures:

- **Carbon Reduction**
  - Reduce the carbon content in stainless steel to below 0.02%, which limits the availability of carbon to form chromium carbide ($Cr_{23}C_6$).
- **Appropriate Heat Treatment**
  - Apply solution annealing followed by rapid cooling (quenching) to retain chromium in solid solution and prevent carbide precipitation.
- **Use of Stabilizing Elements**
  - Add elements such as niobium (Nb) or titanium (Ti), which preferentially form NbC or TiC, consuming excess carbon.
  - This process is known as stabilization, as it prevents Cr depletion by avoiding Cr carbide formation.

Electrochemical Basis for Intergranular Attack

Intergranular corrosion occurs at grain boundaries, which are high-energy regions compared to the grain interiors. This increased reactivity is governed by the free energy relationship:

$$G = -nFE$$

Where:

- G = free energy.
- n = number of electrons,
- F = Faraday's constant.
- E = electrode potential.

Interpretation

- Since free energy is inversely proportional to electrode potential, regions with lower potential (E) have higher G, making them anodic.
- Therefore, grain boundaries act as anodic zones, and the adjacent grain interiors act as cathodes, promoting selective attack along the boundaries.

Surface Etching and Over-etching Effects

- During chemical etching, grain boundaries are attacked first, revealing the grain structure.
- If the surface is over-etched, etch pits may form along dislocation intersections with the surface.
- Dislocation cores possess elevated energy and become anodic sites after grain boundary attack, leading to pit formation and further localized corrosion.

Selective Leaching (Dealloying)

Selective leaching (also known as dealloying) refers to the preferential removal of one element from an alloy, leaving behind a porous and weakened metal structure.

As shown in Fig. 10.10, when an alloy consists of two metals:

- The more active metal dissolves into the solution (e.g., zinc in brass).
- The less active metal remains behind (e.g., copper in brass).

This process results in:

- Mechanical weakening of the material.

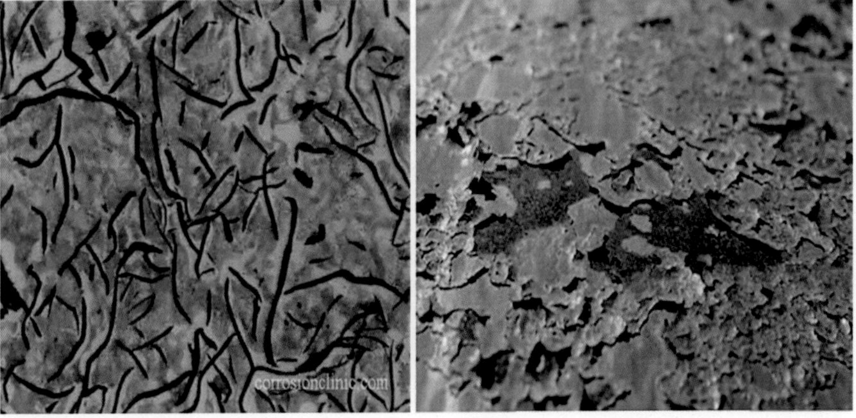

Fig. 10.10 Dealloying

- Porous structure with diminished load-bearing capacity.
- Loss of electrical and thermal conductivity.

Common Forms of Selective Leaching

- Dezincification: Selective removal of zinc from brass.
- Graphitization: Removal of iron from cast iron, leaving behind graphite.
- Denickelification: Removal of nickel from copper-nickel alloys.

Both intergranular corrosion and selective leaching compromise structural integrity through localized mechanisms driven by microstructural or compositional imbalances. Understanding the electrochemical and metallurgical foundations of these phenomena is essential for effective material design and corrosion control.

Dezincification

Dezincification is a type of selective leaching where zinc is preferentially removed from brass alloys, particularly from α-brass (typically 70% Cu, 30% Zn). This form of corrosion leaves behind a porous, weakened copper-rich layer, significantly compromising the mechanical and structural integrity of the material.

Visual and Physical Indicators

- Surface becomes porous and spongy.
- Color change: original yellow surface (from Cu-Zn alloy) turns red due to copper enrichment.
- Often initiated or accelerated by foreign deposits, chloride ions, or stagnant water environments.

Mechanism of Dezincification

The mechanism by which zinc is selectively leached while copper remains has been debated and studied extensively. There are three main theories explaining how the process occurs:

- **Theory 1: Direct Zinc Dissolution**
  - Zinc selectively dissolves from the alloy into the electrolyte as Zn^{2+} ions:

$$Zn \rightarrow Zn^{2+} + 2e^{-}$$

  - Copper remains intact in the metallic lattice.
  - Over time, the surface layer becomes depleted in zinc, leading to structural weakness and copper enrichment.
  - This theory explains the gradual formation of a porous copper matrix.
- **Theory 2: Simultaneous Dissolution and Copper Redeposition**
  - Both zinc and copper dissolve initially from the brass:

$$Cu \rightarrow Cu^{2+} + 2e^{-}$$
$$Zn \rightarrow Zn^{2+} + 2e^{-}$$

- Due to the higher electrochemical nobility of copper, Cu^{2+} ions redeposit back onto the surface:

$$Cu^{2+} + 2e^{-} \rightarrow Cu \left(\text{redeposited}\right)$$

- This results in a copper-rich surface layer, explaining the reddish appearance and porous structure.
- This theory is more widely accepted, as it aligns with observed electrochemical behavior.

- **Theory 3: Combination of Both Mechanisms**
  - In many cases, dezincification involves elements of both mechanisms:

    Zinc selectively dissolves, and copper may partially dissolve and redeposit.
    The extent of each process depends on environmental conditions (e.g., pH, chloride concentration, oxygen availability).

Dezincification leads to copper enrichment and zinc depletion, often resulting in brittle, porous surfaces with reduced mechanical strength. The redeposition of copper and the dissolution of zinc are both critical to the process, which is electrochemically driven by the potential difference between copper and zinc.

Prevention and Control of Dezincification

Dezincification compromises both the mechanical integrity and electrical conductivity of brass components by removing zinc and leaving behind a porous, copper-rich matrix. The following strategies are commonly used to mitigate or prevent this form of selective leaching:

Methods to Prevent or Minimize Dezincification

- Modify Alloy Composition
  - Reduce the zinc content to below 15% to create dezincification-resistant brasses.
  - Alloys with lower zinc concentrations are less prone to selective zinc removal.
- **Environmental Control**
  - Lower the aggressiveness of the environment by:

    Removing dissolved oxygen.
    Reducing chloride or salt concentration.
    Controlling pH and temperature to less aggressive levels.
- **Use of Tin-Modified Brass (Admiralty Brass)**
  - Tin (Sn) additions (typically 1%) result in Admiralty Brass, which offers enhanced resistance to dezincification by forming a more stable and protective passive film.
- **Addition of Inhibiting Elements**
  - Alloying with phosphorus (P), arsenic (As), or antimony (Sb) provides a barrier effect:

These elements redeposit on the surface, blocking active dissolution sites. Their presence inhibits zinc migration, acting as a built-in corrosion inhibitor.

Types of Dezincification
Dezincification can occur in two distinct morphological forms:

- **Layer Type**
  - A uniform, porous copper-rich layer forms on the surface.
  - Often shallow but spread over large areas.
  - Common in stagnant environments or low-velocity fluids.
- **Plug Type**
  - Localized pit-like regions where zinc is removed in isolated pockets, forming copper plugs.
  - These zones are deep and concentrated, potentially initiating crack propagation or leakage paths.

Resulting Consequences
- Both types lead to a significant reduction in mechanical strength and ductility.
- Even visually intact surfaces can be structurally brittle and weakened internally.

Graphitization of Gray Cast Iron
Graphitization is a specific form of selective leaching (dealloying) that affects gray cast iron, where iron is selectively removed, leaving behind a skeleton of graphite flakes. This process significantly deteriorates the mechanical strength of the material while maintaining its outer appearance, often leading to unexpected structural failures.

Metallurgical Interpretation
From a metallurgical standpoint, graphitization involves the decomposition of pearlite (a lamellar mixture of ferrite and cementite) into ferrite and graphitic carbon:

- **Pearlite → Ferrite + Graphite.**

This transformation typically occurs during high-temperature aging or long-term service exposure at elevated temperatures, particularly in environments containing moisture and acidic species.

Electrochemical Mechanism of Graphitic Corrosion
In corrosive environments, galvanic cells form between the graphite flakes (cathode) and the iron matrix (anode):

- Iron dissolves anodically:

$$\mathrm{Fe} \rightarrow \mathrm{Fe}^{2+} + 2\mathrm{e}^{-}$$

- The graphite remains intact, acting as a cathodic site.

This galvanic interaction accelerates the dissolution of iron around the graphite, leading to the progressive loss of metal support and the retention of graphite in place.

Consequences of Graphitization

- The surface may appear intact, but internally, the material becomes porous, brittle, and weak.
- The retained graphite skeleton has little structural strength, resulting in:
  - Loss of tensile and impact strength.
  - Cracking or fracture under mechanical loads.
  - Leakage or collapse in pressure-containing systems.

Graphitization is a destructive form of corrosion in gray cast iron, driven by galvanic action between graphite and iron. While visually deceptive, it severely weakens components and requires careful monitoring and material selection in critical applications.

10.6 Erosion-Corrosion (Wear and Corrosion)

Erosion-corrosion is a form of accelerated attack that results from the combined effect of mechanical erosion (due to high-velocity fluid flow or suspended solids) and electrochemical corrosion. It commonly occurs in components subjected to turbulent or impinging flow, particularly in elbows, reducers, orifices, and heat exchanger tubes, as illustrated in Fig. 10.11.

Mechanism of Erosion-Corrosion

Erosion-corrosion occurs when fluid movement removes the protective surface film (oxide or salt layer), exposing fresh metal to the corrosive environment. The repeated removal and reformation of the passive layer accelerates the corrosion rate.

This process is intensified in the presence of:

- Suspended solids or bubbles.
- Flow-induced vibrations (fretting).
- Galvanic interactions.
- Localized turbulence or pressure drops.

Factors Affecting Erosion-Corrosion

- **Surface Film Properties**
  - Metals with stable and adherent oxide films are more resistant.

    Example:
    Copper forms $CuCl_2$ in NaCl, which is unstable.
    Brass, however, forms a more stable and protective surface, making it more erosion-resistant.
    Copper oxide (CuO) is also adherent, providing some protection.
- **Fluid Velocity**
  - The corrosion rate remains moderate at low velocities, but increases sharply beyond a critical velocity.

Example:
Aluminum in nitric acid shows increased corrosion when velocity exceeds 4 ft/s, due to the removal of protective aluminum nitrate and oxide films (see Fig. 10.11).

- **Turbulence and Impingement**
  - Occurs at:
    Elbows (due to abrupt direction changes).
    Reducers (due to decreased cross-sectional area).
  - Localized turbulence intensifies erosion, often requiring increased wall thickness in these regions as a design mitigation.

Galvanic Effects

- When two dissimilar metals are exposed to a corrosive flow, galvanic corrosion accelerates attack on the anodic metal.
  - Example:

    316 stainless steel coupled with lead (Pb) in high-velocity sulfuric acid (H_2SO_4) results in severe erosion and galvanic degradation.

Material Properties (Metallurgy)

- Materials with:
  - Tough, adherent passive layers.
  - High surface hardness.
  - Alloying elements like molybdenum (Mo) or chromium (Cr)

  offer enhanced resistance.
  - Example:

    18.8 stainless steel (AISI 304) gains better erosion-corrosion resistance with Mo addition.

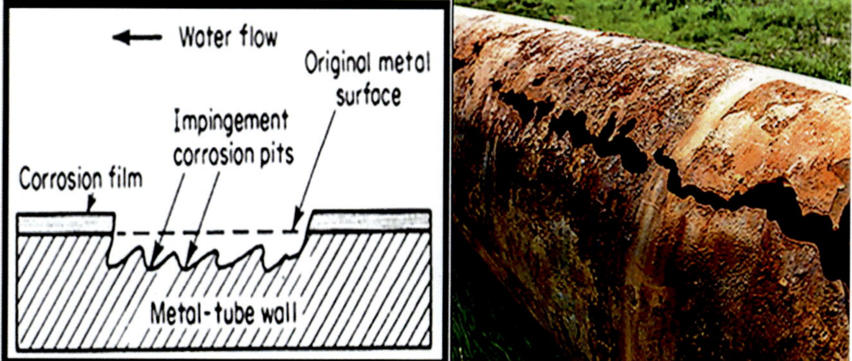

Fig. 10.11 Erosion corrosion

Protection Against Erosion-Corrosion

- Improved Material Selection
  - Use erosion-resistant alloys such as:
    Fe-Cr 80-20 alloy.
    High-Mo stainless steels.
    Ni-based alloys (e.g., Hastelloy).
  - **Design Optimization**
  - Increase thickness in known turbulence-prone regions (e.g., elbows, bends).
  - Use gradual transitions in reducers to reduce turbulence and impingement.
  - Minimize abrupt flow direction changes.
  - **Environmental Modification**
    Filter suspended solids to reduce abrasive erosion.
    Add corrosion inhibitors to minimize anodic activity.
    Remove dissolved oxygen (deaeration) to suppress oxidation reactions.
  - **Protective Coatings**
    Apply hard coatings or epoxy linings in high-velocity zones.
  - **Cathodic Protection**

    Implement sacrificial anode or impressed current systems in buried or submerged structures to counteract corrosion.

Erosion-corrosion is a destructive and often underestimated form of damage in high-flow systems. A multi-faceted approach combining material selection, engineering design, and environmental control is essential to mitigate its effects and extend component life.

10.7 Cavitation Corrosion

Cavitation corrosion is a specialized form of mechanical-chemical degradation that occurs due to the combined effect of fluid pressure variations and electrochemical corrosion. It results from the cyclic formation and violent collapse of vapor or gas bubbles on a metal surface in a corrosive fluid environment.

This phenomenon is commonly observed in systems where high-velocity flow and rapid pressure fluctuations exist, such as in:

- Pump impellers.
- Ship propellers.
- Hydraulic turbines.
- Control valves.

As shown in Fig. 10.12, the repeated implosion of bubbles near the surface creates microjets and shockwaves that:

1. Disrupt and remove protective surface films.
2. Create surface pitting.

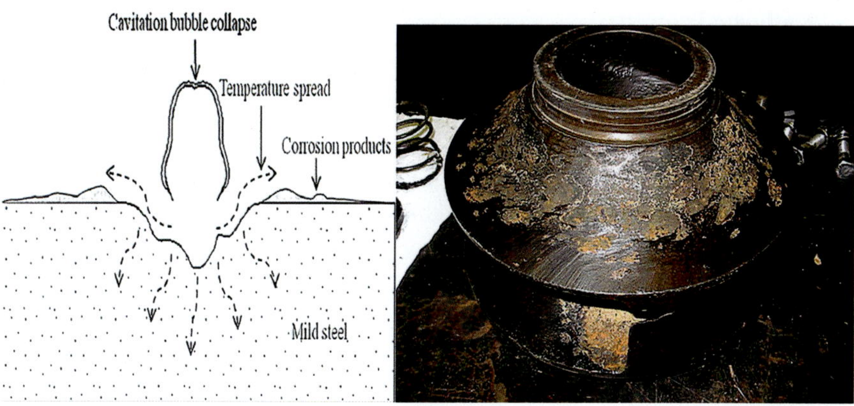

Fig. 10.12 Cavitation damage

3. Expose fresh metal to corrosive media.
4. Accelerate localized corrosion attack.

Mechanism of Cavitation Corrosion

1. Bubble Formation: As liquid pressure falls below the vapor pressure (often at suction or low-pressure zones), vapor bubbles form.
2. Bubble Collapse: When the liquid flows to a higher-pressure region, the bubbles collapse violently, generating localized high pressures and temperatures.
3. Surface Film Disruption: The implosion erodes or strips away the passive film.
4. Accelerated Corrosion: The bare metal is exposed to corrosion, and the cycle repeats.

Control of Cavitation Damage

To minimize or prevent cavitation-related corrosion, the following methods are recommended:

- **Reduce Pressure Fluctuations**
  - Design systems to avoid sudden pressure drops.
  - Maintain steady, uniform flow to prevent vapor formation.
- **Improve Surface Finish**
  - Polish the metal surface to reduce the initiation sites for cavitation pits.
  - A smooth finish minimizes turbulence and bubble nucleation.
- **Use Protective Coatings**
  - Apply hard metallic or polymeric coatings to absorb cavitation impact.
  - Materials like epoxy, nickel plating, or ceramic linings provide durable protection.

- **Cathodic Protection**
  - Helps minimize electrochemical corrosion once the passive film is damaged.
  - Often used in combination with coatings for maximum effect.

Cavitation corrosion is an aggressive, cyclic attack initiated by fluid dynamics and intensified by electrochemical reactions. It is particularly destructive in moving components, and effective control requires both mechanical design optimization and corrosion protection measures.

10.8 Fretting Corrosion

Fretting corrosion is a specialized form of corrosion-erosion that occurs at the contact interface between two metal surfaces subjected to cyclic relative motion while under sustained compressive load or pressure. This interaction causes mechanical wear and oxidation, typically manifesting as pits, grooves, and surface damage accompanied by the accumulation of fine oxide debris, as illustrated in Fig. 10.13.

Conditions for Fretting Corrosion

- Metallic contact under tight pressure.
- Small-scale sliding or vibration between the surfaces.
- Similar material properties of the contacting metals (modulus, hardness).
- Presence of oxygen or reactive species for oxidation.

Common Applications Affected

- Railroad tracks.
- Automotive assemblies.
- Jet engine components.
- Shaft couplings and splines.

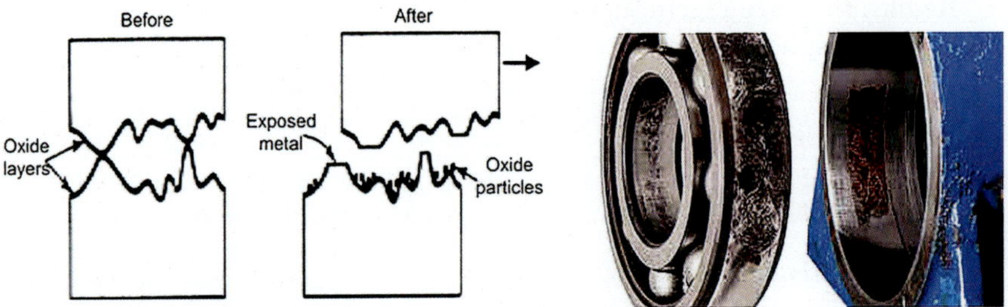

Fig. 10.13 Fretting corrosion

- Bolted and riveted joints under vibration.

Consequences of Fretting
- Initiation of fatigue cracks.
- Loosening of components and dimensional loss.
- Seizing or galling.
- Increased wear and reduction in service life.

Mechanisms of Fretting Corrosion.
1. **Wear–Oxidation Sequence**
    - Initial cold welding occurs at high-pressure micro-contacts.
    - Relative motion breaks the welds, creating grooves or pits.
    - Local temperature rise accelerates oxidation.
    - Oxide debris forms and accumulates in the worn regions, aggravating surface roughness and further wear.
2. **Oxidation–Wear Sequence**
    - The surfaces are pre-oxidized due to exposure.
    - Relative motion disrupts the oxide layer, exposing fresh metal.
    - The cycle of wear and oxidation repeats at the damaged sites.
    - The continual breakdown and reformation of oxides lead to progressive material degradation.

Control and Prevention of Fretting Corrosion
- **Lubrication**
  - Use of anti-wear lubricants or greases to minimize friction and reduce the tendency for adhesion.
- **Surface Coating**
  - Apply protective or hard coatings (e.g., nitride, carbide, or polymer coatings) to increase wear resistance and prevent metal-to-metal bonding.
- **Surface Hardening**
  - Employ **heat treatments** or **alloying** to increase the **hardness** of contact areas, thereby improving resistance to deformation and wear.
- **Dampening Mechanisms**
  - Use of gaskets, springs, or vibration-absorbing materials to limit relative motion and dissipate vibrational energy.

Fretting corrosion is driven by the interplay of pressure, micro-motion, and oxidation, and is particularly dangerous due to its ability to initiate fatigue failures and weaken critical joints. Successful mitigation requires both mechanical design considerations and surface protection strategies.

10.9 Corrosion Cracking

Corrosion cracking is a type of material degradation that arises from the combined action of a corrosive environment and mechanical stress. The stress may be tensile, compressive, or residual, and it facilitates the initiation and propagation of cracks, which are exacerbated by chemical reactions at the crack tip or surface.

This type of failure is insidious, as components may appear intact externally while internal cracking develops progressively.

Categories of Corrosion Cracking

Corrosion cracking is classified into three main forms, depending on the type of stress and chemical environment involved:

- **Stress Corrosion Cracking (SCC)**
  - Occurs when a component is exposed to a specific corrosive environment and subjected to constant tensile stress.
  - The cracks initiate and propagate at stress levels below the material's yield strength.
  - SCC is often intergranular or transgranular, depending on the alloy and environment.
  - Critical factors:

    Material susceptibility.
    Presence of tensile stress.
    Specific corrosive medium (e.g., Cl^- for stainless steels, NH_4NO_3 for copper alloys).

Example: Austenitic stainless steel exposed to chlorides under tensile load.

- **Corrosion Fatigue (CF)**
  - Results from the combined effects of cyclic mechanical loading and corrosive attack.
  - Corrosion accelerates the initiation of fatigue cracks and reduces the fatigue life of the material.
  - Cracks initiate at surface defects and propagate more rapidly than in air due to localized corrosion at the crack tip.

Example: Aircraft components or rotating shafts in seawater environments.

- **Hydrogen-Induced Cracking (HIC)**

Also known as Hydrogen Embrittlement (HE)

- Caused by hydrogen atoms diffusing into the metal during corrosion, welding, or cathodic protection.
- Tensile stress facilitates hydrogen accumulation at microstructural defects, leading to brittle cracking.
- High-strength steels and titanium alloys are particularly vulnerable.
- Cracking can be delayed, occurring hours or even days after exposure.

Example: Cracking in pipeline steels exposed to sour gas (H_2S environments).

Corrosion cracking poses a serious reliability risk across industries, especially in structural and pressure-bearing components. Preventive strategies include material selection, stress relief, environmental control, and protective coatings.

10.9.1 Stress Corrosion Cracking (SCC)

Stress corrosion cracking (SCC) refers to the unexpected failure of metals or alloys caused by the simultaneous action of a corrosive environment and static tensile stress. The stress level at which SCC occurs is often much lower than the material's yield strength, typically around 60% of the design or yield stress.

SCC can occur in aqueous solutions, humid environments, or gaseous atmospheres, and is particularly dangerous because it can cause catastrophic failure without significant plastic deformation or warning signs.

Key Features of SCC

1. Pure metals are generally immune or resistant to SCC; alloys are more susceptible.
2. Requires static tensile stress (applied or residual).
3. It depends strongly on the chemical nature of the environment.
4. Anodic polarization (positive current) accelerates SCC; cathodic polarization (negative current) delays or prevents it.
5. The crack growth rate in SCC lies between mechanical fatigue and corrosion alone.
6. Fracture can be intergranular (along grain boundaries), transgranular (through grains), or a combination of both.

Mechanisms of Stress Corrosion Cracking (SCC)

Stress corrosion cracking (SCC) can be broadly classified into two primary categories based on the underlying mechanisms.

1. **Dissolution-Based Mechanisms**

   These mechanisms involve localized metal dissolution at the crack tip, driven by the combined effects of tensile stress and a corrosive environment.

   - **Pre-existing Active Path Model**
     - In sensitized 18-8 austenitic stainless steels, chromium carbide precipitation leads to chromium-depleted grain boundaries. These regions act as anodic paths that promote preferential dissolution. Under tensile stress, cracks initiate and propagate along these weakened zones (see Fig. 10.14).

2. **Strain-Generated Active Path Mechanisms**

   These mechanisms arise from deformation-induced processes that generate active dissolution sites during mechanical loading.

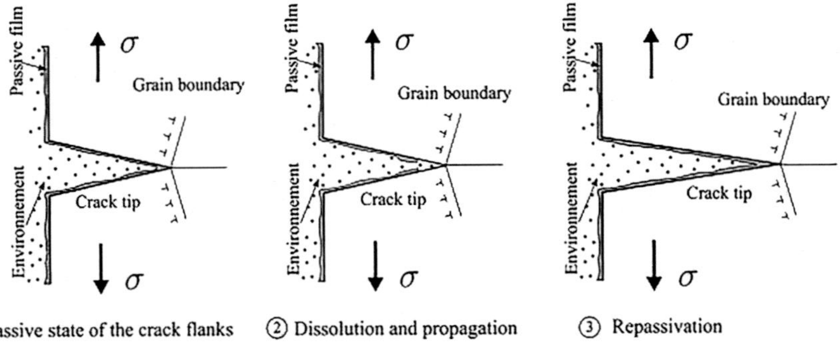

Fig. 10.14 Strain generation active path (film rupture model)

- **Film Rupture Model**
  - A passive film forms at the crack tip; however, under tensile stress, this film ruptures, exposing fresh metal to the corrosive environment and resulting in localized dissolution.
  - Crack propagation proceeds through cyclic processes governed by:
  - the rate of re-passivation the rate of re-passivation,
  - the rate of dissolution, and
  - slip-step dissolution.
- **Slip-Step Dissolution Model**
  - Dislocation movement results in the formation of slip bands on the metal surface. At these sites, freshly exposed metal undergoes repeated cycles of dissolution and re-passivation, enabling progressive crack growth (see Fig. 10.15).

Corrosion Tunnel Model

- Localized dissolution under stress creates corrosion tunnels (see Fig. 10.16).
- As tunnels expand, the remaining ligaments thin and can fail mechanically, leading to ductile fracture.

2. **Cleavage-Based Mechanisms**

   These mechanisms emphasize embrittlement at the crack tip, often due to chemical species such as hydrogen.

   - **Adsorption-Induced Cleavage**
     - Corrosive species infiltrate the lattice, weakening interatomic bonds and promoting brittle cleavage.

   Tarnish Rupture Model (see Fig. 10.17)
     - A passive film breaks under stress, exposing fresh metal.
     - A new film forms, halting crack growth until it ruptures again.
     - Crack arrest marks form due to the intermittent nature of propagation.

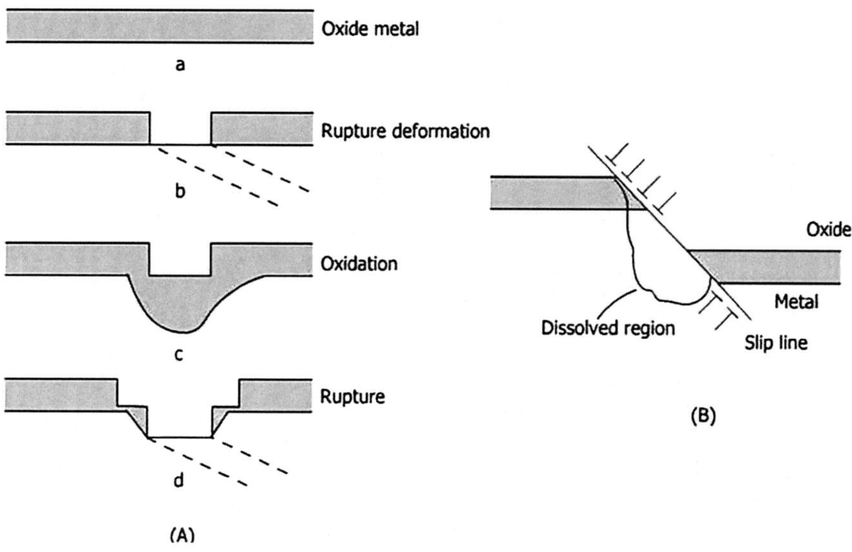

Fig. 10.15 Strain generation active path (slip-step dislocation model)

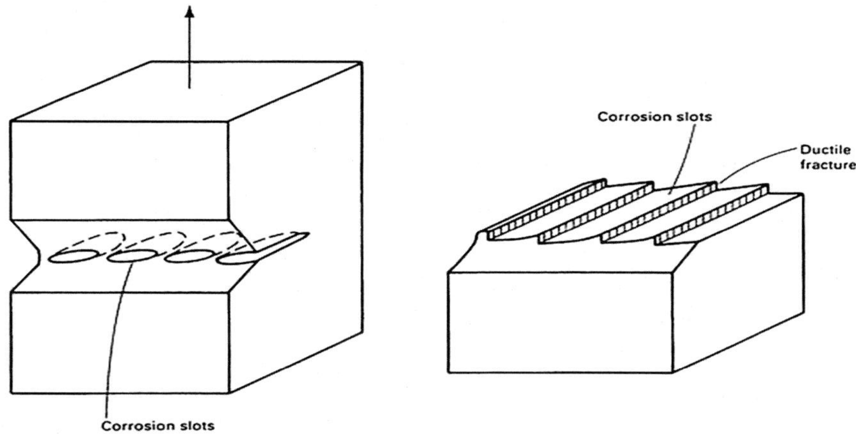

Fig. 10.16 Strain generation active path (corrosion tunnel model)

Film-Induced Cleavage (see Fig. 10.18)

- Similar to tarnish rupture, but the film forms from metal dissolution rather than oxidation.
- Seen in Cu-Zn or Cu-Al alloys, especially where a de-zincified layer forms.

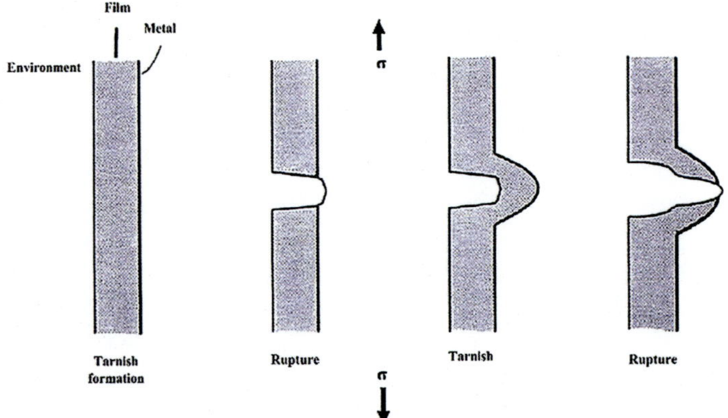

Fig. 10.17 Tarnish rupture

Fig. 10.18 Film-induced cleavage

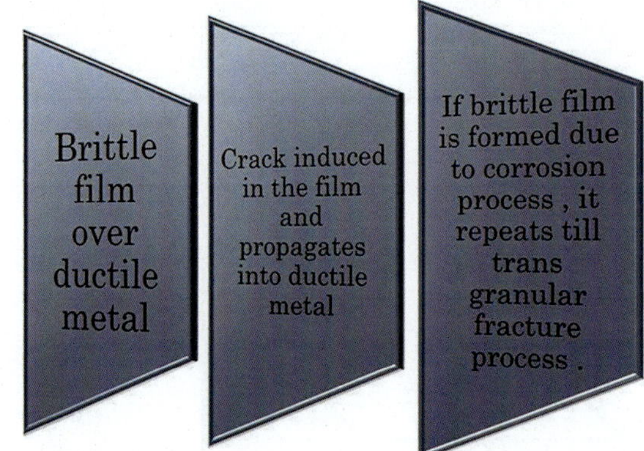

- **Atomic Surface Mobility**
  - Crack growth proceeds as atoms diffuse away from the crack tip.
  - This occurs more readily with surface-active elements or low-melting-point compounds, especially in high-temperature or nitric acid environments.

SCC Threshold Stress and Failure

As shown in Fig. 10.19, SCC does not occur below a threshold stress level, which depends on:

- Material/alloy.
- Environment (solution, pH, temperature).
- Stress concentration.

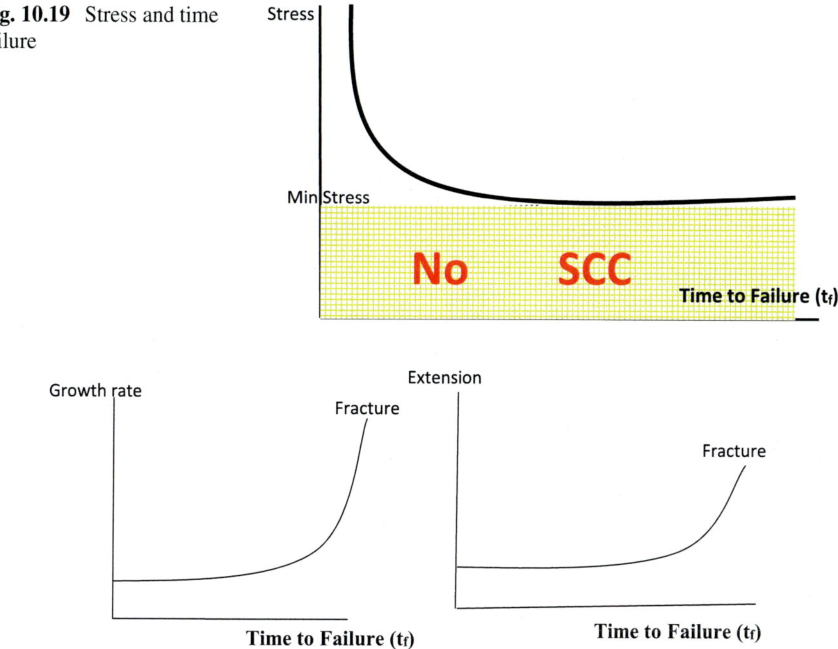

Fig. 10.19 Stress and time failure

Fig. 10.20 Growth rate and extension and time to failure

Crack growth can be **slow initially**, then **accelerates rapidly** near failure (Fig. 10.20).

Environmental Influences

- Temperature increases diffusivity and crack propagation rate.
- pH affects passivation and dissolution rates.
- Oxygen presence may promote SCC in alloys such as 304 stainless steel in NaCl.

Prevention and Control of SCC

1. Reduce tensile stress, especially residual stress, via annealing.
2. Remove aggressive species, such as oxygen or chlorides.
3. Material substitution with SCC-resistant alloys.
4. Cathodic protection, especially for aqueous systems.
5. Inhibitor addition, such as oxygen scavengers.

10.10 Hydrogen-Related Cracking

Hydrogen-related cracking encompasses various failure modes caused by the ingress and interaction of hydrogen with metals and alloys. These failures are particularly severe in high-strength steels and hydride-forming metals such as titanium and zirconium. Hydrogen atoms are typically introduced during cathodic reactions, welding, or corrosive exposure.

The main mechanisms include:

- Hydrogen Embrittlement.
- Decohesion.
- Hydrogen-Induced Plasticity.
- Hydrogen Blistering.

Hydrogen Embrittlement

Hydrogen embrittlement occurs when atomic hydrogen diffuses into a metal under tensile stress, particularly at crack tips. In hydride-forming materials (e.g., Zr, Ti), the hydrogen combines with the metal to form brittle hydrides. These brittle phases reduce ductility and facilitate crack propagation.

- Source of hydrogen: Cathodic reaction.
- Effect: Repeated hydrogen absorption and hydride formation lead to progressive crack growth and eventual failure.

Decohesion

In the decohesion mechanism, hydrogen reduces the interatomic bond strength, making fracture possible at much lower stress levels. This process is explained by Griffith's fracture criterion, where hydrogen reduces the surface energy (γs), and hence, the critical fracture stress (σc):

$$\sigma_c = \sqrt{\frac{2E\gamma_s}{\pi c}}$$

Where:

- σ_c: Critical fracture stress,
- E: Modulus of elasticity.
- γ_s: Surface energy,
- c: Half-length of the existing crack.

As γ_s decreases due to hydrogen, the material becomes more prone to brittle fracture.

Hydrogen-Induced Plasticity

In this mechanism, hydrogen reduces the resistance to dislocation movement at the crack tip, allowing plastic deformation to occur more easily. This leads to:

- Reduced strain hardening.
- Enhanced crack propagation under lower applied stresses.

This is particularly critical in ductile metals where plasticity normally hinders crack growth.

Hydrogen Blistering
Hydrogen blistering results from the accumulation of atomic hydrogen in internal voids. The hydrogen atoms recombine into molecules (H_2) within these voids, generating internal pressures exceeding 100 atm. This pressure causes:

- Crack initiation and propagation.
- Formation of visible blisters on the surface.

The blisters compromise both structural integrity and surface appearance.

Prevention of Hydrogen-Related Cracking
1. Avoid internal voids or porosity in the material.
2. Use inhibitors to suppress atomic hydrogen formation.
3. Eliminate hydride-forming elements (e.g., Zr, Ti) where possible.
4. Minimize hydrogen-generating reactions, such as excessive cathodic polarization.

10.11 Corrosion Fatigue CF

Corrosion fatigue (CF) is the combined action of cyclic mechanical loading and a corrosive environment, leading to the premature failure of materials. Unlike fatigue in an inert environment, the presence of corrosion removes the fatigue limit, accelerating crack initiation and growth.

Fatigue Behavior With and Without Corrosion
- Without corrosion, steel exhibits a fatigue limit—below which fatigue failure does not occur, regardless of the number of cycles.
- With corrosion, this limit disappears; fatigue failure can occur at any stress level.
- Nonferrous metals, such as aluminum and copper alloys, do not have a fatigue limit, even in non-corrosive environments. Every stress level corresponds to a finite fatigue life, which decreases further in corrosive conditions.

Crack Appearance in Fatigue and CF
As shown in Fig. 10.21, the material fails under alternating tensile or compressive stress even when the applied stress is below the yield strength.

Figure 10.22 illustrates the **surface appearance** of fatigue failure:

- **Without corrosion**:
  - The smooth region indicates fatigue crack initiation and propagation.
  - The rough region appears where the remaining uncracked section can no longer withstand the load, leading to sudden fracture (overload failure).

- **With corrosion**:
  - Corrosive species infiltrate the crack tip, promoting localized attack and accelerated crack growth.

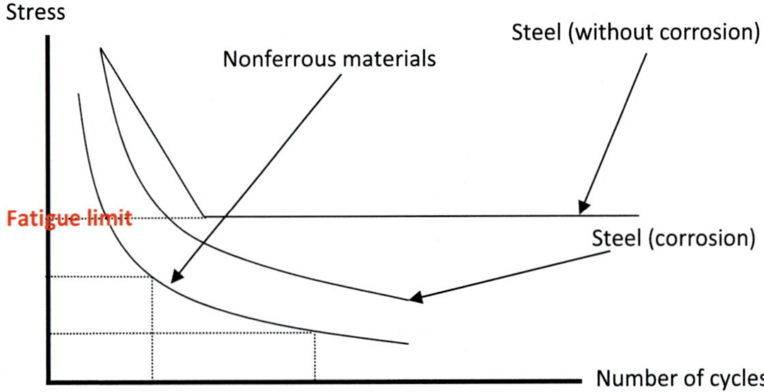

Fig. 10.21 Corrosion fatigue for steel and nonferrous

Fig. 10.22 Fatigue fracture

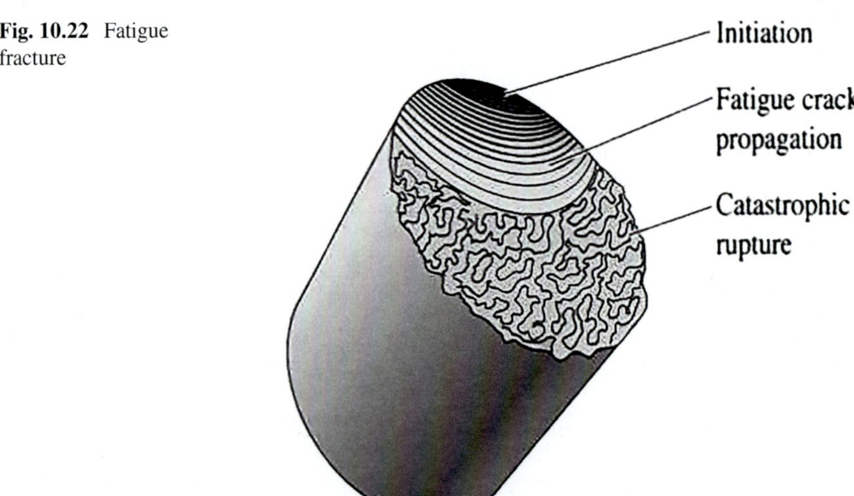

– The entire fracture surface becomes rough due to the formation of corrosion products and the irregular nature of the crack path.

Prevention of Corrosion Fatigue

To mitigate the risk of corrosion fatigue, the following strategies are recommended:

1. **Material Selection**
   Use materials with high corrosion resistance and fatigue strength, such as stainless steels or corrosion-resistant alloys.

2. **Surface Protection**
   Apply coatings, anodizing, or cathodic protection to limit corrosion at crack initiation sites.
3. **Environmental Control**
   Prevent conditions that promote localized corrosion:
   - Control oxygen content.
   - Maintain neutral or controlled pH.
   - Minimize temperature fluctuations.
   - Reduce the concentration of aggressive ions, such as chlorides.
4. **Design Improvements**
   Avoid features that create stress concentrations, such as sharp corners or sudden changes in cross-section, and eliminate crevices or pits that act as corrosion initiation sites.

Preventing Fatigue Failure Due to CF
- Material
- Avoid the presence of pitting or crevice corrosion by controlling environmental factors such as:
  - Oxygen content.
  - Temperature.
  - pH.
  - Concentration of corrosive species.

10.12 Biologically Influenced Corrosion (BIC)

Biologically influenced corrosion (BIC), also known as microbiologically influenced corrosion (MIC), is a type of corrosion where the presence and activity of biological organisms—either microorganisms or larger biofouling organisms—influence the corrosion behavior of metals. These organisms are commonly found in soil, groundwater, oil emulsions, cutting fluids, and water systems (industrial, domestic, or marine).

Effects on Corrosion Behavior
- Direct effects: Bioactivity can alter anodic or cathodic reactions.
- Indirect effects: Biological activity may produce by-products (e.g., acids, sulfides) that modify surface films, reduce pH, or deplete oxygen, accelerating corrosion.

1. **Microbiological Corrosion**
   Microbiological corrosion is typically localized, often resembling pitting or crevice corrosion. It is primarily caused by two types of bacteria:
   (a) **Anaerobic Bacteria (absence of oxygen)**

One of the most common anaerobic bacteria is the sulfate-reducing bacterium (SRB), which thrives in wet, clay-rich, deaerated environments, such as waterlogged soils or stagnant water during the rainy season.

Key reactions:

- Sulfate reduction by SRB

$$SO_4^{2-} + 8H^+ \rightarrow S^{2-} + 4H_2O$$

- Hydrogen evolution (cathodic reaction)

$$2H^+ + 2e^- \rightarrow H_2$$

- Metal dissolution and sulfide formation

$$Fe \rightarrow Fe^{2+} + 2e^-$$

$$Fe^{2+} + S^{2-} \rightarrow FeS(\text{black precipitate})$$

Iron sulfide (FeS) on a metal surface is an indication of SRB activity.

(b) **Aerobic Bacteria (presence of oxygen)**

In the presence of air, sulfur-oxidizing bacteria convert sulfur into sulfuric acid, which aggressively attacks metal surfaces:

$$2S + 3O_2 + 2H_2O \rightarrow 2H_2SO_4$$

- pH decreases, enhancing metal dissolution.
- Slime layers formed by microbial colonies reduce oxygen availability beneath them, promoting the growth of anaerobic bacteria, creating a microenvironment for corrosion.

2. **Macrofouling**

Macrofouling organisms, including barnacles, mollusks, and snails, readily attach to submerged metallic surfaces in marine environments. Their presence contributes to corrosion through several mechanisms:

(a) The accumulation of biological deposits promotes differential aeration, leading to localized corrosion such as crevice corrosion.
(b) Increased surface roughness and drag on structures (e.g., ship hulls), resulting in higher fuel consumption and reduced efficiency.
(c) Formation of anaerobic microenvironments beneath deposits, which facilitate microbial activity and accelerate localized corrosion processes.

Prevention of Biologically Influenced Corrosion (BIC)

To mitigate BIC, several strategies can be implemented:

1. **Aeration**
   - Enhancing oxygen availability helps suppress the growth of anaerobic microorganisms responsible for corrosion.
2. **Chlorination and Biocides**
   - Chemical treatment of water systems is employed to eliminate or inhibit microbial populations.
3. **Protective Coatings**
   - Application of tar, enamel, or polymer-based coatings provides a physical barrier that isolates the metal surface from the environment.
4. **Cathodic Protection**
   - Corrosion is reduced by shifting the electrochemical potential of the metal into a protective (cathodic) range.
5. **Material Selection and Substitution**
   - Use of non-metallic materials (e.g., plastics or asbestos cement pipes) is recommended in environments prone to biological activity.
6. **Maintenance and Cleaning**
   - Regular removal of biofilms and deposits prevents the development of localized corrosive microenvironments.

10.13 Liquid Metal Attack or Embrittlement (LME)

Liquid metal embrittlement (LME) refers to a phenomenon in which a solid metal becomes brittle and fractures upon contact with a liquid metal, even at relatively low temperatures and stress levels. This interaction may cause sudden failure, reduction in ductility, or intergranular penetration, depending on environmental and material conditions.

Modes of Failure Due to LME

1. Instantaneous failure: Sudden fracture of the solid metal when under applied tensile or residual stress.
2. Delayed failure: Failure occurs below the metal's normal tensile strength, often after an incubation time.
3. Intergranular penetration: Liquid metal can infiltrate the grain boundaries of the solid metal without requiring applied stress.
4. High-temperature corrosion: In some cases, elevated temperature accelerates metal–metal reactivity, intensifying LME effects.

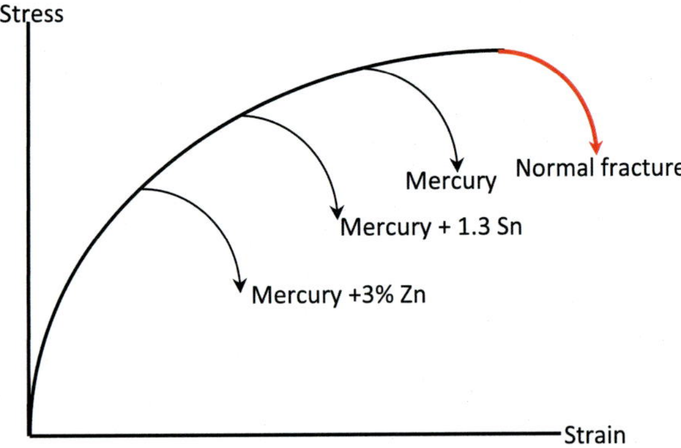

Fig. 10.23 Effect of liquid metal on the engineering stress–strain curve

Consequences of LME

1. Loss of ductility.
2. Decrease in fracture strength and strain.
3. Ductile-to-brittle transition:
    - Intergranular fracture, e.g., brass exposed to mercury.
    - Transgranular cleavage, depending on the metal system.
4. **Environment-specific behavior**:
    - **Example:** Zn in liquid Hg can cause instantaneous fracture in susceptible alloys.
5. **Severity of embrittlement**:
    - As shown in Fig. 10.23, the stress-strain curve of a solid metal (e.g., polycrystalline aluminum) shifts dramatically when exposed to a liquid metal like mercury. Certain alloying additions can reduce this effect.

Factors Affecting Embrittlement Severity

- Chemical reactivity of the liquid and solid metals.
- Type of alloying elements.
- Grain size of the solid metal.
- Presence of micro-cracks or stress concentrators.

Requirements for LME to Occur

1. Intimate contact: Good wetting between liquid and solid metal surfaces.
2. Sufficient stress: Applied or residual tensile stress to induce plastic deformation.
3. Adequate supply of liquid metal in contact with the solid.

Influencing Factors (Table 10.1)

Mechanisms of Liquid Metal Embrittlement

1. **Reduction in surface energy**:
    - Liquid metal lowers the crack tip surface energy, decreasing fracture stress.

Table 10.1 Conclusion of influencing factors and their description

Factor	Description
Grain size	Coarse-grained materials are generally more susceptible to liquid metal embrittlement (LME) than fine-grained materials
Temperature	LME occurs when the solid metal is exposed to temperatures above the melting point of the liquid metal
Strain rate	Higher strain rates typically increase the susceptibility to LME
Alloying elements	Alloying elements may either increase or decrease LME susceptibility, depending on their solubility and interaction with the liquid metal

2. **Stress-assisted dissolution**:
   - Preferential dissolution of atoms at the crack tip under stress promotes crack propagation.
3. **Adsorption-induced weakening**:
   - Adsorption of metal atoms reduces atomic cohesion, particularly at grain boundaries.
4. **Reduction in tensile or shear strength**:
   - The liquid metal interferes with atomic bonding, promoting brittle fracture under stress.

Preventive Measures Against LME

1. Impurity additions: Introducing certain impurity atoms into the solid metal may reduce wettability and inhibit embrittlement.
2. Solute additions in the liquid metal: For example, adding lead (Pb) to bismuth (Bi) reduces embrittlement effects in copper.
3. Use of barriers: Applying a ceramic coating between liquid and solid metal prevents direct interaction.
4. Cladding techniques: Cladding zircaloy with pure zirconium helps reduce embrittlement from liquid cadmium.
5. Stress management: Reducing or relieving residual and applied tensile stresses lowers the likelihood of LME initiation.

Chapter 11
Corrosion Protection

11.1 Change of the Materials

Material Selection and Composition Adjustment

Corrosion resistance can be significantly improved by selecting appropriate materials or modifying their composition. One effective approach is purifying metals by removing harmful impurities. For instance, eliminating sulfur or phosphorus from steel enhances its corrosion resistance. Similarly, reducing carbon content in stainless steel increases its resistance to intergranular corrosion.

Alloying is another key strategy. Adding elements such as antimony (Sb) and arsenic (As) to brass acts as an inhibitor, protecting it against de-alloying and enhancing its resistance. In stainless steel, alloying with chromium (>12%) and nickel (Ni) improves its corrosion-resistant properties. The addition of molybdenum (Mo) further enhances resistance, especially in chloride-containing environments. Titanium (Ti) and niobium (Nb) act as carbon stabilizers, preventing chromium carbide formation at grain boundaries and thus mitigating sensitization.

The addition of noble metals such as palladium (Pd) or platinum (Pt) to titanium enhances its corrosion resistance and improves passivation. These elements serve as cathodic sites, accelerating the formation of a stable passive film.

Microstructural Modification

Heat treatment and mechanical working can alter the microstructure to improve corrosion resistance. For example, desensitizing stainless steel involves heating to 1000 °C followed by rapid cooling, which dissolves carbides and restores uniform chromium distribution. This eliminates anodic and cathodic sites along grain boundaries and prevents intergranular corrosion.

Residual Stress Relief

Residual tensile stress can lead to stress corrosion cracking (SCC). Applying low-temperature annealing can relieve these stresses and reduce the likelihood of SCC.

M. M. Ghatus, *Fundamentals of Corrosion Science and Engineering*,
https://doi.org/10.1007/978-3-032-13138-6_11

Surface Compression (Shot Peening)
Introducing compressive stress on the surface via shot peening can significantly reduce the susceptibility to fatigue corrosion and improve fatigue resistance.

Material Selection Guidelines
The corrosion potential (equilibrium potential) of a metal or alloy guides its suitability for specific environments:

- **E < −0.414 V**: Corrosion occurs in both neutral and acidic environments without oxygen.
- **−0.414 V ≤ E < 0.0 V**: Corrosion occurs in acidic media without oxygen and in neutral media with oxygen.
- **0.0 V ≤ E < 0.815 V**: Corrosion occurs in both acidic and neutral media in the presence of oxygen.
- **E > 0.815 V**: Metals typically exhibit high corrosion resistance, though they may still corrode in acidic environments with oxygen.

Passivation Tendency
Some metals, although thermodynamically active (low equilibrium potential), can form protective passive films, significantly improving corrosion resistance. Titanium exhibits the highest tendency to passivate, followed by aluminum and chromium.

The *coefficient of passivity* quantifies the ease of passivation and is defined by the equation:

$$\text{Coffecient of passivity} = \frac{E_{corr} - E_a}{E_C - E_{corr}}$$

Where:

- E_a = anodic potential.
- E_c = cathodic potential.
- E_{corr} = corrosion potential.

Formation of Protective Insoluble Products
Some metals resist corrosion by forming insoluble protective films. For example, pure lead (Pb) resists dilute sulfuric acid by forming a $PbSO_4$ layer. However, in concentrated acid, $PbSO_4$ dissolves and corrosion increases.

Effects of Impurities and Alloying Additions

- Zinc in HCl: High-purity zinc has good resistance. However, the addition of iron (Fe) or copper (Cu) significantly increases the corrosion rate due to galvanic effects (Fig. 11.1).
- Iron and Carbon: In iron, carbon increases corrosion susceptibility through the formation of iron carbide (Fe_3C).
- Titanium and Noble Metals: Adding Pt or Pd to Ti promotes passivation, reducing the corrosion rate.

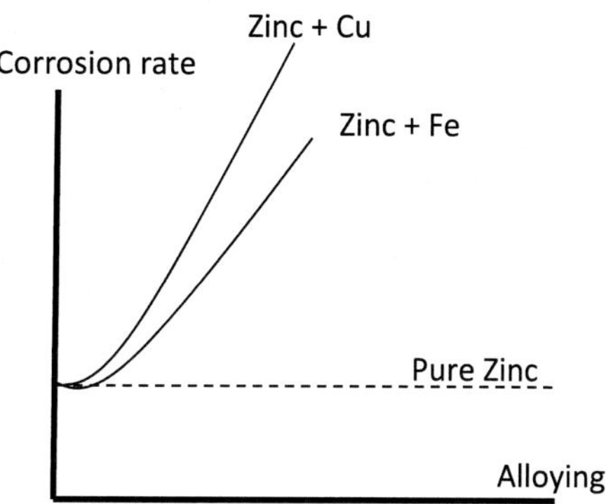

Fig. 11.1 Corrosion rate and alloying of Zn with Fe

- Zinc and Cadmium: Cadmium additions to zinc increase cathodic polarization, reducing corrosion current (Fig. 11.2).
- Magnesium and Manganese: In Mg alloys exposed to 3% NaCl, Mn additions decrease corrosion rates and enhance resistance.
- Iron and Manganese: Mn mitigates the negative effects of iron and increases hydrogen overvoltage, thereby decreasing the corrosion rate (Fig. 11.3).

11.2 Corrosion Control by Design

Effective corrosion prevention begins with proper design considerations that account for environmental and operational factors. Key principles include:

1. **Appropriate Site Selection**
   Factors such as prevailing wind direction, ambient salinity, and optimized stacking arrangements can significantly influence corrosion behavior.
2. **Plant Layout**
   A layout that minimizes stagnant zones and maximizes accessibility for inspection and maintenance can reduce corrosion risk.
3. **Material Selection**
   Choosing materials compatible with the operating environment and the type of corrosion anticipated is essential.
4. **Component Design**
   Designing components with corrosion resistance in mind can significantly extend service life.
5. **Protection Measures**
   Incorporating corrosion prevention strategies, such as coatings or cathodic protection, should be planned during the design stage.

Key Design Rules for Minimizing Corrosion

1. **Increase Material Thickness**

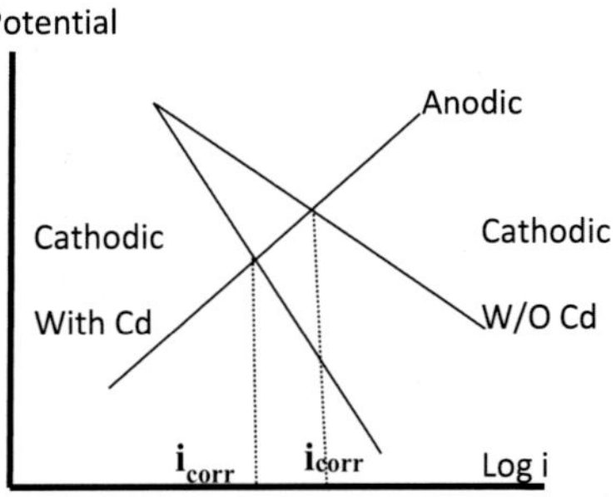

Fig. 11.2 Effect of Cd on corrosion current

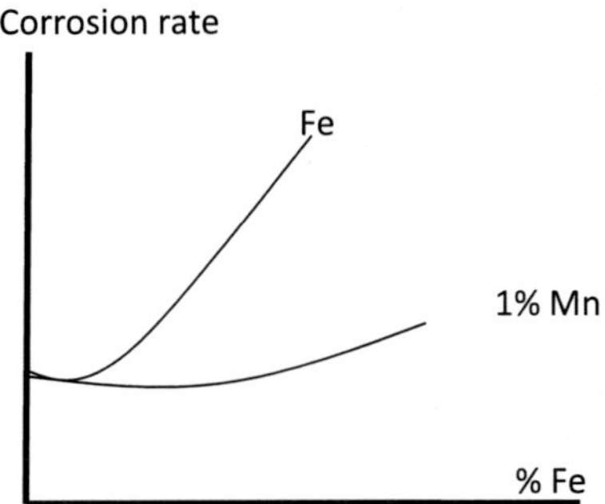

Fig. 11.3 Corrosion rate and alloying of Fe with Mn

Thicker sections provide a longer corrosion allowance and increased equipment lifetime under harsh conditions, especially at elevated temperatures and pressures.

2. **Avoid Galvanic Contact**
   Prevent electrical contact between dissimilar metals to minimize galvanic corrosion.
3. **Eliminate Crevice Formation**
   Design joints and assemblies to avoid narrow, stagnant areas where crevice corrosion may initiate.
4. **Ensure Proper Drainage**
   Avoid liquid traps in tanks and piping by incorporating sloped surfaces or drain holes.
5. **Avoid Sharp Corners**
   Design smooth transitions to reduce turbulence in fluid flow, minimizing erosion and corrosion.
6. **Minimize Mechanical Stress Concentrations**
   Avoid sharp notches or cracks that may act as stress raisers and initiate stress corrosion cracking (SCC).
7. **Prevent Vapor Pockets**
   Avoid designs that allow gas accumulation at the top of vessels. Use vacuum pumps or venting systems when necessary.
8. **Exclude Air/Oxygen**
   Reduce oxygen content in closed systems to suppress cathodic reactions and decrease the corrosion rate.

11.3 Change of Environment

Altering the environment surrounding a metal structure is one of the most effective ways to control corrosion. Techniques include:

- **Moisture Removal**

- Use dehumidifiers or desiccants such as silica gel to reduce moisture content in enclosed environments.
- **Humidity Control**
- Maintain the temperature above the dew point (typically around 6–7 °C above ambient) to minimize condensation.
- **Oxygen Removal**
- Eliminating dissolved oxygen reduces cathodic reactions and slows corrosion. This can be achieved by:
  - Inert gas purging using nitrogen or argon.
  - Vacuum evacuation.
  - Chemical scavengers:
    Sodium sulfite (Na_2SO_3) at low temperatures.
    Hydrazine (N_2H_4) at high temperatures, which reacts with oxygen to form nitrogen and water.
- **Acid Removal/pH Adjustment**
- Adjusting the pH to a neutral or slightly alkaline range decreases acidic corrosion risk.
- **Chloride Ion Removal**
- Chloride ions are aggressive and disrupt passive films. Their removal is critical, especially for stainless steels and aluminum.
- **Solid Particle Removal**
- Filters or separators should be used to remove particulates that may cause crevice corrosion or erosion.
- **Salt Removal**
- Desalination techniques reduce electrolyte conductivity and thereby decrease corrosion rates.
- **Inhibitor Addition**

- Corrosion inhibitors are chemicals added in low concentrations (typically in parts per million) to reduce the aggressiveness of the medium by forming protective films or slowing electrochemical reactions.

Chapter 12
Inhibitors

In corrosion science, an inhibitor is a substance that decreases the corrosion rate or completely stops a chemical reaction or a corrosion process.

Characteristics of Inhibitors

1. The trial and error method for the choice of inhibitors in a particular situation.
2. Organic or inorganic.
3. Costly and toxic.
4. Specific.
5. Synergistic in nature (two or more inhibitors).
6. Increasing temperature, increasing chlorine ions, and decreasing the pH all increase the required inhibitors.

12.1 Types of Inhibitors

In corrosion science, an *inhibitor* is a chemical substance that reduces or entirely halts the corrosion process by interfering with the electrochemical reactions occurring at the metal surface.

Characteristics of Corrosion Inhibitors

1. Empirical Selection: The selection of inhibitors is often based on a trial-and-error method tailored to specific environments.
2. Composition: Inhibitors may be either organic or inorganic.
3. Cost and Toxicity: Many are expensive and potentially hazardous.
4. Specificity: Most inhibitors are effective only under certain conditions or with particular metals.
5. Synergistic Behavior: Two or more inhibitors can act synergistically to enhance performance.

M. M. Ghatus, *Fundamentals of Corrosion Science and Engineering*,
https://doi.org/10.1007/978-3-032-13138-6_12

6. Environmental Sensitivity: Increased temperature, higher chloride concentrations, and lower pH levels demand higher concentrations or more robust inhibitors.

1. **Passivators (Anodic Inhibitors)**

   These inhibitors reduce corrosion by stabilizing or enhancing the passive film on a metal surface. Their effectiveness depends on maintaining concentrations above a critical value to sustain passive current density (as shown in Fig. 12.1 in the log *i* vs. *potential* diagram).

   Benefits:

   - Stabilize and maintain the passive oxide layer.
   - Promote repassivation.
   - Form insoluble, protective surface films that repair damage.

   Types:

   - **Direct Passivators**: Strong oxidizing agents such as chromate (CrO_4^{2-}) and nitrite (NO_2^-). These do not require the presence of oxygen to induce passivation.
   - **Indirect Passivators**: Substances like sodium hydroxide (NaOH) that enhance oxygen adsorption by consuming H^+ ions, thus suppressing the cathodic reaction and improving corrosion resistance.

2. **Barrier-Type Inhibitors**

   These inhibitors form a protective, insoluble layer (barrier) on the metal surface.

   - **Example**: Zn^{2+} ions in neutral media react with OH^- to form zinc hydroxide:

$$2OH^- + Zn^{2+} \rightarrow Zn(OH)_2$$

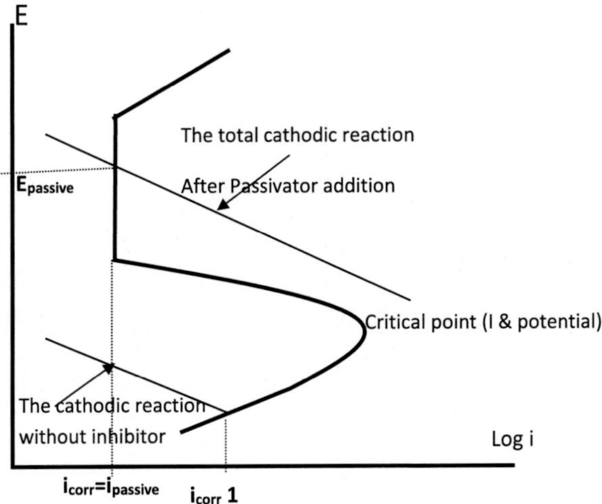

Fig. 12.1 Effect of inhibitor on the corrosion rate of active–passive metal

Depending on how the barrier acts, these may function as anodic or cathodic inhibitors.

3. **Organic Adsorption Inhibitors**

   Typically used in oil and gas environments, these inhibitors (e.g., mercaptans or sulfides) adsorb onto the metal surface, forming a hydrophobic oily layer that:

   - Blocks hydrogen ion adsorption.
   - Prevents metal ion solvation. This reduces both anodic and cathodic reactions.

4. **Inorganic Barrier Inhibitors**

   These form insoluble precipitates with metal ions to provide protection.

   - **Example**: Bicarbonate (HCO_3^-) forms iron bicarbonate ($Fe(HCO_3)_2$) on iron surfaces in alkaline media, acting as a corrosion barrier.

5. **Poisons (Cathodic Inhibitors)**

   These special cathodic inhibitors work in acidic media to:

   - Prevent hydrogen gas evolution by inhibiting the hydrogen recombination step.
   - Suppress oxygen reduction.

   Examples:

   Phosphorus (P), arsenic (As), and antimony (Sb) are added to brass (~1%) to prevent dezincification. Their mechanism includes:

   - Cu^{2+} ions re-deposit on the surface.
   - OH^- reacting with Sb to form $Sb(OH)_4^-$.
   - The formation of a surface film at cathodic sites that inhibits further corrosion.

6. **Vapor Phase Inhibitors (VPIs)**

   These inhibitors volatilize and form a protective layer on metal surfaces in enclosed environments.

   - **Example**: Dicyclohexylamine nitrate (DCHN), with a vapor pressure of 3×10^{-3} atm at 25 °C, is used in electronic components to form a passivating film on internal metal parts.

Effectiveness of Inhibitors (Based on Mixed Potential Theory)

Figure 12.2 illustrates the application of mixed potential theory to corrosion inhibitors:

- **Cathodic Inhibitors**: Shift the cathodic polarization curve downward, reducing the corrosion current.
- **Anodic Inhibitors**: Shift the anodic curve to reduce corrosion.
- **Combined Use**: The simultaneous use of both inhibitors provides the greatest reduction in corrosion current, offering enhanced protection.

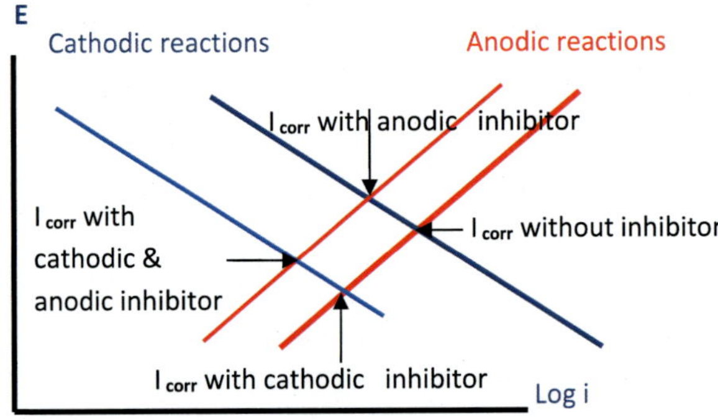

Fig. 12.2 The mixed potential theory and inhibitors

Chapter 13
Protective Coating

Protective coatings are primarily applied to improve corrosion and wear resistance on metal surfaces. Additionally, they can enhance fatigue performance and extend the lifespan of components in corrosive environments.

13.1 Benefits of Protective Coatings

1. Increase corrosion resistance.
2. Act as a barrier between the metal and the corrosive medium.
3. Provide cathodic protection (e.g., zinc coatings on iron or galvanized coatings).
4. Function as inhibiting layers by releasing protective compounds.
5. Improve electrical insulation by increasing surface resistance.

13.2 Key Characteristics of a Good Coating

1. Excellent corrosion resistance over time.
2. Strong adhesion to the substrate.
3. Continuity (free of defects such as holidays or pinholes).
4. Long service life.
5. Proper surface preparation before application.

Note Surface preparation is crucial and varies depending on the coating type. It includes removing contaminants like oil, grease, or oxide layers to ensure proper adhesion and performance.

M. M. Ghatus, *Fundamentals of Corrosion Science and Engineering*,
https://doi.org/10.1007/978-3-032-13138-6_13

13.3 Types of Coatings

1. **Metallic Coatings**
   - **Anodic Coatings**: Sacrificial coatings that protect the underlying metal, such as zinc (Zn) on iron (Fe).
   - **Cathodic Coatings**: Applied noble metals, such as tin (Sn) or gold (Au) on steel. These require defect-free application; otherwise, the unfavorable area ratio (large cathode, small anode) accelerates corrosion at the defect.
   - **Advantages**:
     - High durability.
     - Decorative appeal.
     - Effective shielding against environmental factors.
2. Non-Metallic Coatings
   - **Vitreous Enamel**: A hard, glassy coating offering chemical resistance and durability.
   - **Portland Cement**: A thick, alkaline coating offering protection in high-pH environments.
   - **Chemical Conversion Coatings**: Formed by chemical interaction between the metal surface and treatment agents:
     - ***Examples***: Anodizing (Al), chromate conversion, phosphate, oxide layers, nitriding, carburizing, rust converters.
   - **Organic Coatings**: Includes paints, resins, lacquers, and plastic-based films.
     - Multi-layer systems are often used to avoid pinholes and ensure coverage.
     - **Common Failures**:

     Blistering.
     Premature corrosion.
     Delamination.
     Filiform corrosion (a form of localized corrosion under painted surfaces).

13.4 Methods of Application

13.4.1 Mechanical Methods

- **Cladding**: Bonding a corrosion-resistant metal to a less-resistant base.
- **Mechanical Plating**: Cold application of metal powders.
- **Painting**: Application of organic coatings with brushes, rollers, or sprays.

13.4.2 Physical Methods

- **High-Temperature Techniques**:
  - Hot dipping (e.g., zinc galvanizing).
  - Welding overlays.
  - Thermal spray fusing.
  - Laser-based coating.
- **Low-Temperature Techniques**:
  - Cathodic sputtering.
  - Metal spraying.

13.4.3 Chemical Methods

- **High-Temperature Techniques**:
  - Electroless (autocatalytic) plating.
  - Chemical vapor deposition (CVD).
  - Aluminizing (diffusion-based).
- **Low-Temperature Techniques**:
  - Electroplating (e.g., Ni, Cr, Cu).
  - Anodizing (especially aluminum).
  - Phosphating (e.g., zinc phosphate coatings).
  - Vacuum deposition (e.g., thin metal films).

Chapter 14
Oxidation and Hot Corrosion Phenomenon

Oxidation, often referred to as hot corrosion, is the reaction between a metal and oxygen at elevated temperatures. This process can lead to the formation of oxide layers on the metal surface, which may either protect or degrade the metal depending on the nature of the oxide formed.

14.1 Pilling–Bedworth Ratio (PBR)

One of the first scientific criteria for evaluating a material's oxidation resistance is the **Pilling–Bedworth Ratio (PBR)**. This ratio helps predict whether the oxide layer formed during oxidation will be protective or non-protective.

P_{BR} Formula:

$$R_{PB} = \frac{W/D}{w/d}$$

Where (Table 14.1)**:**

- R_{PB} = Pilling–Bedworth Ratio.
- w = Molecular weight of the metal,
- d = Density of the metal,
- W = Molecular weight of the oxide.
- D = Density of the oxide.

An ideal oxide layer should be compact, adherent, and continuous, which typically occurs when RPBR_{PB}RPB lies between **1 and 2** (Table 14.2).

For example, the oxidation of aluminum:

M. M. Ghatus, *Fundamentals of Corrosion Science and Engineering*, https://doi.org/10.1007/978-3-032-13138-6_14

Table 14.1 Interpretation of the Pilling–Bedworth Ratio, oxide behavior, and examples

PBR range	Oxide behavior	Example metals
$R_{PB} < 1$	The oxide layer is **too thin** and non-protective (may crack)	Magnesium, potassium
$1 < R_{PB} < 2$	The oxide layer is **protective**, passivating, and adherent	Aluminum, titanium, chromium
$R_{PB} > 2$	The oxide layer is **too thick**, prone to **spalling or flaking**	Iron (Fe_2O_3), silicon

Table 14.2 Values of R_{PB} for different materials

Metal	Metal oxide	Formula	R_{PB}
Potassium	Potassium oxide	K_2O	0.474
Sodium	Sodium oxide	Na_2O	0.541
Lithium	Lithium oxide	Li_2O	0.567
Strontium	Strontium oxide	SrO	0.611
Calcium	Calcium oxide	CaO	0.64
Barium	Barium oxide	BaO	0.67
Magnesium	Magnesium oxide	MgO	0.81
Aluminum	Aluminum oxide	Al_2O_3	1.28
Lead	Lead(II) oxide	PbO	1.28
Platinum	Platinum(II) oxide	PtO	1.56
Zirconium	Zirconium(IV) oxide	ZrO_2	1.56
Zinc	Zinc oxide	ZnO	1.58
Hafnium	Hafnium(IV) oxide	HfO_2	1.62
Nickel	Nickel(II) oxide	NiO	1.65
Iron	Iron(II) oxide	FeO	1.7
Titanium	Titanium(IV) oxide	TiO_2	1.73
Iron	Iron(II,III) oxide	Fe_3O_4	1.90
Chromium	Chromium(III) oxide	Cr_2O_3	2.07
Iron	Iron(III) oxide	Fe_2O_3	2.14
Silicon	Silicon dioxide	SiO_2	2.15
Tantalum	Tantalum(V) oxide	Ta_2O_5	2.47
Niobium	Niobium pentoxide	Nb_2O_5	2.69
Vanadium	Vanadium(V) oxide	V_2O_5	3.25
Tungsten	Tungsten(VI) oxide	WO_3	3.3

$$Al + (3/2)O_2 \rightarrow (1/2)\ Al_2O_3$$

Where:

$W = 101.96$ g, $D = 3.19$ g/cm^3, $w = 26.98$ d = 2.7 g/cm^3

$$R_{PB} = \frac{\dfrac{101.96}{2 * 3.19}}{\dfrac{26.98}{2.7}} = 1.3$$

The reported value is 1.28, ≈ 1.3.

Another example is the oxidation of chromium

$$Cr + (3/2)O_2 \rightarrow (1/2)\ Cr_2O_3$$

$W = 152$ g $D = 5.21$ g/cm^3, $w = 52$ g d = 7.14 g/cm^3

$$R_{PB} = \frac{\dfrac{152}{2 * 5.210}}{\dfrac{52}{7.14}} = 2.0029$$

The reported value is 1.99, ≈ 2.0029.

Oxidation of Fe

$$Fe + (1/2)\ O_2 \rightarrow FeO$$

$W = 55$ g $D = 5.754$ g/cm^3, $w = 55.15$ g $d = 7.86$ g/cm^3

$$R_{PB} = \frac{\dfrac{55.85}{5.745}}{\dfrac{55.15}{7.86}} = 1.76$$

The reported value is 1.77, ≈ 1.76.

Summary of Your Statement (With Clarification)

When **1 mole of a metal** oxidizes:

- **If the molar volume of the oxide > molar volume of the metal**

→ The oxide layer is **protective**, covers the entire surface, and prevents further oxidation.

→ **PBR > 1** → Good corrosion resistance (e.g., Aluminum, Chromium, Titanium).

- **If the molar volume of the oxide < molar volume of the metal**

→ The oxide layer is **non-protective**, cracks, or exposes bare metal, allowing continued corrosion.
→ **PBR < 1** → Poor corrosion resistance (e.g., Magnesium, Potassium).

- **If the PBR is much greater than 2**

→ The oxide layer is too bulky and stresses itself, causing it to **flake off or spall**, leading again to continued corrosion.
→ **PBR > 2** → Poor protection despite a thick layer (e.g., Iron(III) oxide on steel).

1. **Thermal Expansion Compatibility**
   - Why it matters: During heating and cooling, if the metal and oxide expand or contract at different rates, mechanical stresses build up.
   - Result: These stresses can crack or spall the oxide layer.
   - Good example: Al and Al_2O_3 have similar expansion coefficients → stable oxide.
   - Bad example: Iron and Fe_2O_3 mismatch → oxide layer often flakes off.
2. **Volatility of the Oxide**
   - Evaporation of oxides like V_2O_5 and WO_3 removes the protective layer.
   - High PBR (>3) can also indicate excessive oxide growth that leads to instability and spalling.
   - Result: The protective barrier is lost, and the bare metal is re-exposed.
3. **Plastic Deformability of Oxide**
   - **Why it matters:** If the oxide can deform plastically (like the metal does under stress), it can stay **adherent** and **intact**.
   - **If not:** Brittle oxides crack under stress → accelerated oxidation.
4. **Low Diffusion Coefficients**
   - O^{2-} or metal cation diffusion through the oxide is the primary mechanism of oxide growth.
   - If diffusion is slow → oxide grows slowly and remains dense and protective.
   - If diffusion is fast → oxide becomes porous or uneven, failing to block further oxidation.
   - Example: Al_2O_3 and Cr_2O_3 have very low oxygen/metal ion diffusivity, so they form excellent barriers.

14.1.1 Sequence of Events During Oxidation

1. **Adsorption of Oxygen on the Metal Surface**
   Oxygen molecules from the environment are adsorbed onto the clean metal surface. At high temperatures, oxygen dissociates into atoms upon contact.

2. **Initial Oxide Nucleation and Lateral Growth**
   Once thermodynamically favorable, oxygen atoms react with metal atoms at discrete sites on the surface to form metal oxide (MOx). These oxide nuclei grow laterally across the surface, attempting to form a continuous protective layer. Simultaneously, oxygen may dissolve into the metal, especially in early stages.
3. **Perpendicular (Thickness-Wise) Growth of the Oxide Layer**
   After coalescence into a continuous layer, the oxide grows in thickness. This growth is governed by:
   (a) Cation diffusion from the metal outward to react with oxygen.
   (b) Anion (oxygen) diffusion inward to react with the metal.
   (c) Electron conductivity across the oxide layer to maintain charge neutrality.
4. **Formation of Defects and Porosity**
   As oxidation proceeds, lattice mismatches, stress from growth, and differential ion movement lead to:
   (a) Vacancies and interstitials.
   (b) Voids and micro-pores in the oxide layer.
5. **Crack Initiation and Propagation**
   Continued growth and thermal cycling can cause:
   (a) **Macro-crack formation** in the oxide layer.
   (b) **Spallation** or detachment of the oxide layer.
   (c) **Increased diffusion** of oxygen and metal ions through cracks, accelerating oxidation.
6. **Special Degradation Phenomena**
   (a) **Evaporation of Oxides**
   Certain oxides (e.g., M_o, O_3, V_2, O_5, WO_3) have low boiling points and may evaporate, leaving bare metal exposed.
   (b) **Internal Oxidation and Embrittlement**
   Oxygen can penetrate along grain boundaries, leading to:
   - **Internal oxidation.**
   - **Grain boundary weakening and embrittlement.**

14.2 Thermodynamic of Oxidation

A chemical reaction at a particular temperature and pressure is related to the free energy. When the free energy (ΔG) is negative (less than zero), a spontaneous reaction is possible; for example:

$$2\mathrm{M}(\mathrm{s})+\mathrm{O}_2(\mathrm{g})\rightarrow 2\mathrm{MO}(\mathrm{s}) \text{ at } (T,P=1\,\mathrm{atm})$$

$$\Delta G = \Delta G^{\circ} + RT \ln K$$

$$K = \frac{a_{MO}^2}{a_M^2 P_{O2}} = \frac{1}{P_{o2}}$$

$$\Delta G = \Delta G^{\circ} + RT \ln \frac{1}{p_{o2}}$$

$$\Delta G = \Delta G^{\circ} - RT \ln p_{o2}$$

K is the equilibrium constant in the nonstandard condition, R is the gas constant, and T is the temperature.

At the equilibrium condition $\Delta G = 0$

$$\Delta G^{\circ}_{T,p} = RT \ln p_{o2}$$

Suppose the partial pressure of oxygen at the equilibrium condition at (T) is less than the actual partial pressure that the metal is exposed to. The metal oxide formation will be possible because the $\Delta G_{(T,P)}$ is less than zero (−ve).

$$\Delta G = RT \ln p_{o2}(\text{equi}) - RT \ln p_{o2}(\text{actual})$$

If the $RT \ln p_{o2}$(actual) > $RT \ln p_{o2}$(equi).
$\Delta G < 0$ oxidation is possible
But, if $RT \ln p_{o2}$(actual) < $RT \ln p_{o2}$(equi)
$\Delta G > 0$ oxidation is not possible, and dissociation of metal oxide will happen
But, when

$$RT \ln p_{o2}(\text{actual}) = RT \ln p_{o2}(\text{equi})$$

ΔG = zero, the process is in an equilibrium condition.

$$\Delta G^{\circ}_{T,P} = RT \ln p_{o2}$$

The dissociation pressure of oxide is equal to:

$$P_{O2}^{\text{dissociation}} = \exp \frac{\Delta G^{\circ}}{RT}$$

In addition, ΔG° can be calculated by knowing the standard enthalpy and entropy change

$$\Delta G_0(T) = \Delta H_0(T) - T \, \Delta \, S_0(T)$$

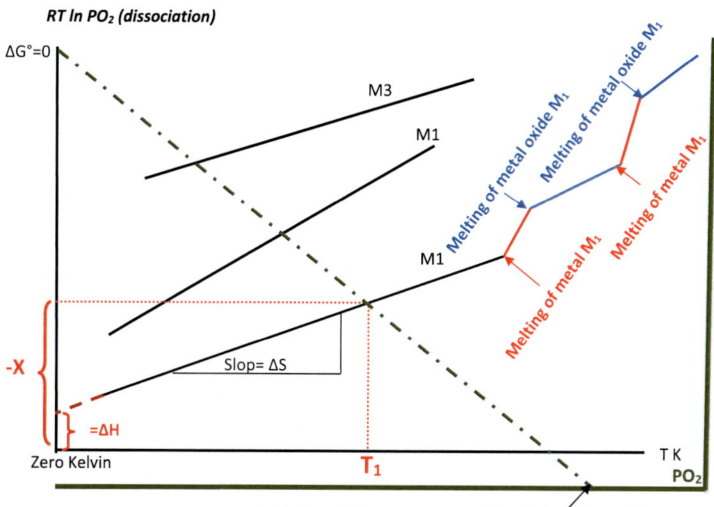

Fig. 14.1 Ellingham diagram

$$RT \ln p_{o2}^{\text{dissociation}} = \Delta H_0 (T) - T \, \Delta S_0 (T)$$

14.2.1 Ellingham Diagram

The Ellingham diagram Fig. 14.1 represents the relationship between $RT \ln Po_2$ and temperature (T) and predicts the equilibrium temperature between a metal, its oxide, and oxygen.

Ellingham's diagram illustrates the behavior of many metals, M_1, M_2, and M_3, and their oxides (MnO). The relationship is linear. Therefore, the slope equals the standard entropy change (ΔS_0). The intersection with the Y-axis is the value of the change of the enthalpy (ΔH_0) when the temperature is equal to zero Kelvin.

At any temperature (T_1) for any metal oxide, the dissociation pressure of oxygen at this temperature can be determined as shown for the M_1.

From the Ellingham diagram, when temperature is known, such as (T_1), the dissociation pressure of oxygen can be calculated by using the following equation:

$$-X = RT_1 \ln p_{O2} (\text{Dissociation})$$

The slope changes in M_1 mean the melting phenomenon of the metal or its oxide is taking place. When the entropy and the slope increase, the metal's melting occurs. Nevertheless, when the slope and entropy decrease, the metal oxide melts. This change is shown in the Ellingham diagram of M_1

14.3 Oxidation Kinetics

Kinetics decides whether the oxidation happens or not. The laws of oxidation kinetics are:

1. Linear law

$$\int \frac{dw}{dt} = K_l$$

$$\Delta w = K_1 t + C$$

Where Δw = weight gains/area, K_1 is the linear rate constant (oxidation) at temperature T, C constant, and t = time

Examples of the linear law are sodium (Na) and potassium (K). The R-value (Pilling–Bedworth Ratio) for the linear law (Na and K) is less than one ($R_{PB} < 1$). This means the metal oxide volume is insufficient to cover the entire surface; therefore, some of the metal surface is still exposed to the oxygen.

In the case of niobium, the Pilling–Bedworth Ratio is greater than 1 ($R_{PB} \gg 1$) $R_{PB} \approx 2.5$. Therefore, the volume of metal oxide is very high, and compressive stress may be generated on the oxide layer, which may cause the oxide layer to crack. This crack will allow oxygen to get into the metal surface.

When the Pilling–Bedworth Ratio equals 1 or is close to 1 ($R = 1$), metal oxide provides good oxidation resistance, for example, Al in an atmospheric environment where $R = 1.28$.

2. Parabolic oxidation law.

Most metals and engineering alloys follow parabolic kinetics at elevated temperatures; the diffusion of ions or electrons usually governs the oxide growth process through the initially formed oxide scale. The rate of weight gain is inversely proportional to the thickness.

$$\int \frac{dw}{dt} = \frac{1}{x}$$

$$\int \frac{\rho A dx}{dt} = \frac{k}{x}$$

$$\rho A \int \frac{dx}{dt} = \frac{k}{x}$$

$$\rho A \int x dx = k \int dt$$

$$\rho A \frac{x^2}{2} = k_p t + c$$

ΔW is the weight gain per unit area; ρ is constant, so

$$\Delta W^2 = 2k_p t + c$$

The square change in weight per unit oxide area is proportional to time. As time increases, the oxidation rate gradually decreases, and the rate of weight gain per unit time also decreases.

The oxidation rate and the weight gain happen at a specific temperature, such as T_1. As shown in Fig. 14.2, if the temperature changes to T_2 ($T_2 < T_1$), the oxidation rate and the weight gain will decrease when the temperature is decreased.

The activation energy (Q) can be calculated by the exposure of the metal for a period at a constant temperature (T_1). The weight gained (Δw) is calculated by weighing the sample before and after exposure. The exact process is repeated at different temperatures, T_2 and T_3, respectively ($T_3 < T_2$), and the final chart is produced as illustrated in Fig. 14.2. The square weight gained with time is shown in Fig. 14.3. A linear relationship is produced between time and the square weight gained.

The aim of Fig. 14.3 is to find the parabolic rate constant (k_p) for every temperature, which is the slope of the line ($\Delta w^2/t$). From the three parabolic rate constants, the activation energy can be calculated as below:

Arrhenius Equation

$$k_p = k_o \exp\left(\frac{-Q}{RT}\right)$$

$$\ln k_p = \ln k_o - \left(\frac{Q}{RT}\right)$$

The plot in Fig. 14.4 shows the relationship between the $\ln k_p$ and ($1/T$); the slope of the straight line is equal to ($-Q/R$), and the value of R is the gas constant. Therefore, the activation energy is easy to calculate by knowing the slope.

3. Logarithmic oxidation law

$$\Delta W = k_{\log} \log t + A$$

Where $k_{\log}$ is the logarithmic oxidation rate constant, t is the time, ΔW is the weight gain per unit area, and A is a constant.

For example, Al and Cu at ambient or slightly higher ambient temperatures obey a Logarithmic oxidation law.

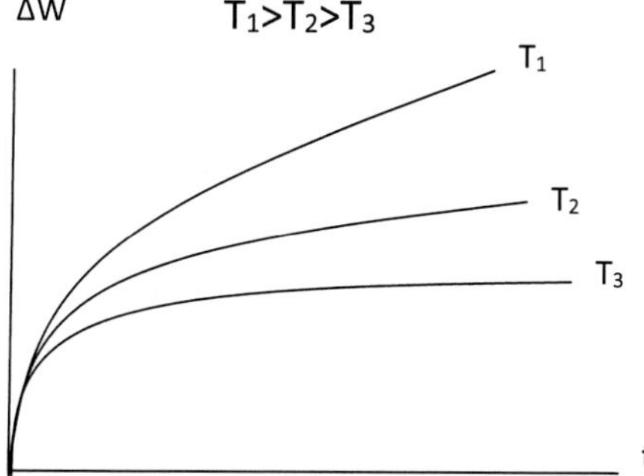

Fig. 14.2 Weight gain rate and temperature effect

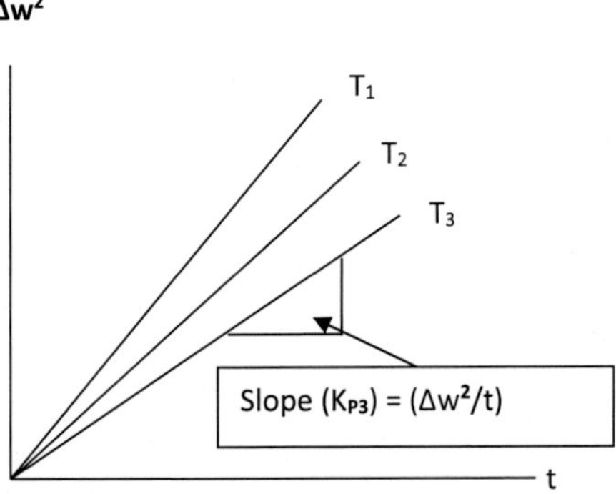

Fig. 14.3 Square weight gain rate and temperature effect

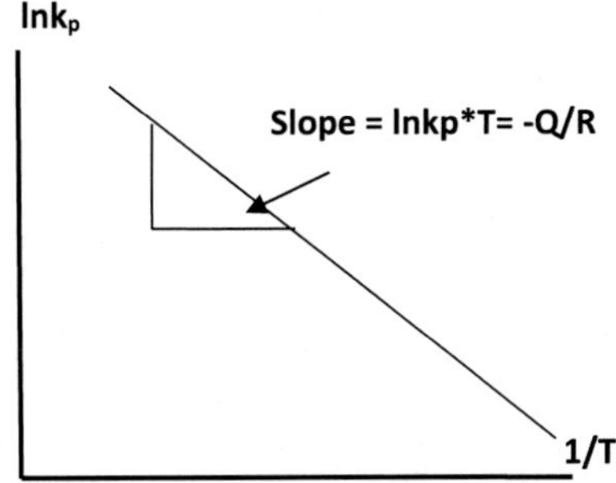

Fig. 14.4 ln k_p and $(1/T)$

4. Cubic oxidation law.

$$\Delta w^3 = k_{cubic} + D$$

For example, Zr follows the cubic oxidation law under a particular condition when two processes simultaneously happen:

- Diffusion-controlled growth of the oxide layer.
- Simultaneous dissolution of oxygen in metal.

A comparison between all the mentioned laws is concluded in Fig. 14.5.

14.3.1 Defect Structure of Metal Oxide and Oxidation

When a metal is oxidized at the metal/metal oxide interface, it loses electrons and forms metal ions through an anodic reaction. At the metal oxide/environment interface, these electrons are consumed in a cathodic reaction, where they combine with oxygen to form oxygen anions. These anions then react with the metal ions to produce metal oxide. In this process, the metal oxide layer functions as both a conductor and an electrolyte.

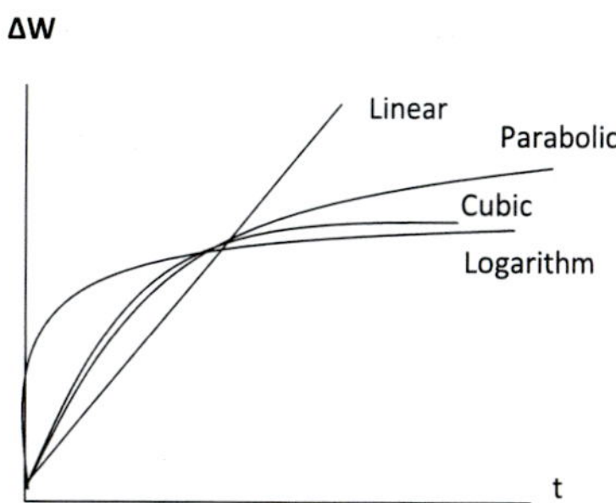

Fig. 14.5 Comparison between all laws of oxidation

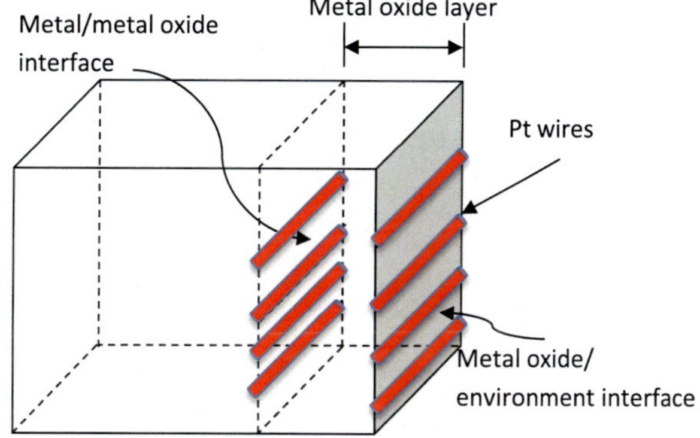

Fig. 14.6 Marker experiment

The growth of the metal oxide layer can occur in one or both of the following ways:

- At the oxide/environment interface (cathodic surface): Metal ions migrate outward and react with oxygen anions.
- At the metal/oxide interface (anodic surface): Oxygen anions diffuse inward through the oxide and react with metal ions.

Marker Experiment

A marker experiment (Fig. 14.6) is employed to determine whether metal oxide growth occurs at the metal/metal oxide interface or the metal oxide/environment interface. Platinum (Pt) wires are typically used as inert markers. These wires must have a diameter smaller than the oxide layer's thickness. The effectiveness of the experiment relies on the inertness of the marker (Pt), ensuring it does not react with either the metal or the oxide at elevated temperatures..

After the installation of platinum (Pt) wires on the metal surface, the metal is subjected to high-temperature oxidation. The position of the Pt wires after oxidation reveals the dominant direction of ion diffusion and the location of oxide growth:

- If the Pt wires are located at the outer interface (closer to the environment), this indicates inward diffusion of oxygen, meaning that oxidation and oxide growth occurred at the metal/metal oxide interface.

- If the Pt wires are located at the inner interface (closer to the metal), this indicates outward diffusion of metal cations, and oxide growth occurred at the metal oxide/environment interface, where oxygen is available.

Typically, both inward and outward diffusion can occur, and the movement takes place through defects in the oxide layer.

Defects in Solids

Point defects play a crucial role in the oxidation process. These defects are thermodynamically stable and include vacancies and interstitial atoms (either cations or anions). Other types of defects include line defects (e.g., dislocations), surface defects (e.g., grain boundaries), and volume defects (e.g., voids). However, point defects are of particular interest in the context of oxidation.

The number of point defects in a solid can be calculated using the formula:

$$n = A_N \exp\left(\frac{-\Delta H_v}{RT}\right)$$

Where:

- n: is the number of defects,
- A_N: is Avogadro's number,
- ΔH_v is the molar energy for defect formation,
- R: is the gas constant,
- T is the temperature in Kelvin.

In ionic oxides, electroneutrality is maintained by balancing the charges of cations and anions. Point defects in oxides can be stoichiometric (with a 1:1 ratio of anions to cations, as in MO) or non-stoichiometric, such as (Fig. 14.7):

- Schottky defects: where both a cation and an anion are missing, creating two vacancies.

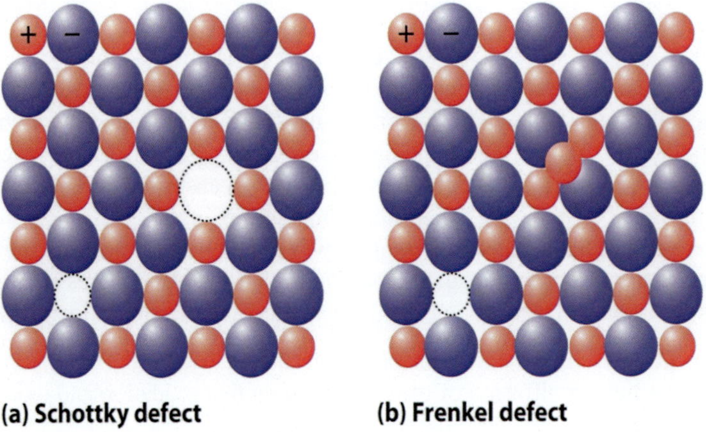

Fig. 14.7 Stoichiometric defect

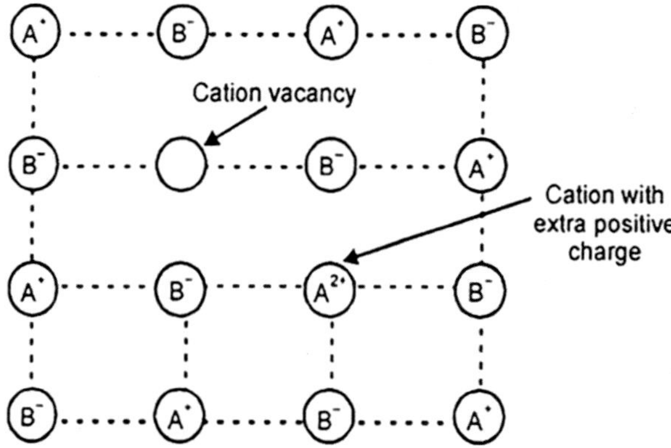

Fig. 14.8 Nonstoichiometric defect

- Frenkel defects: where a cation leaves its lattice site and moves to an interstitial site, maintaining overall stoichiometry.

Figure 14.8 illustrates nonstoichiometric defects, which arise when the ratio of cations to anions deviates from the ideal 1:1 stoichiometry. These defects influence the type of charge carriers and the diffusion mechanisms in oxides:

- In P-type oxide: The dominant charge carriers are *electron–hole pairs.*, Cation vacancies are the primary defect, allowing diffusion occurs through these vacant cation sites.
- In N-type oxide: The dominant charge carriers are *electrons*. This type of oxide typically exhibits excess metal ions or oxygen vacancies, contributing to enhanced electron conductivity.

Chapter 15
Hot Corrosion

Hot corrosion is considered a degradation process rather than a typical corrosion process because it involves the transformation of a previously formed protective oxide layer into a non-protective scale due to reactions with environmental contaminants. For instance, in a gas turbine, air is compressed and mixed with fuel, and combustion raises the temperature to approximately 700–1400 °C. If sulfur (S) is present in the fuel, it reacts with the protective oxide on turbine components, leading to its breakdown and accelerating the degradation of the underlying material.

Turbines are often operated in marine environments where sodium chloride (NaCl) is prevalent. In such settings, the combined presence of sulfur (S) from the fuel, NaCl, oxygen, and moisture can lead to complex chemical interactions. These species may react to form sodium sulfate (Na_2SO_4) or other corrosive compounds, initiating or accelerating hot corrosion through reactions such as:

$$\mathrm{NaCl} + \mathrm{SO_3} + \tfrac{1}{2}\mathrm{O_2} + \mathrm{H_2O} \rightarrow \mathrm{Na_2SO_4} + 2\mathrm{HCl}$$

This reaction leads to the formation of acidic species (e.g., HCl) and sulfates that degrade the protective oxide layer, resulting in the formation of porous, non-adherent scales that cannot prevent further oxidation or metal loss.

Breakdown of Sodium Sulfate (Na_2SO_4):

$$\mathrm{Na_2SO_4} \rightarrow \mathrm{Na_2O}\ (\text{basic oxide}) + \mathrm{SO_3}\ (\text{acidic oxide})$$

These two components aggressively react with the metal oxide layer (MO) that would normally serve as a corrosion-resistant barrier.

M. M. Ghatus, *Fundamentals of Corrosion Science and Engineering*,
https://doi.org/10.1007/978-3-032-13138-6_15

15.1 Types of Hot Corrosion

- **Type I Hot Corrosion—High-Temperature Hot Corrosion**
  - Temperature range: ~850 to 950 °C.
  - Cause: Occurs due to the formation of molten sodium sulfate (Na_2SO_4) on the metal surface.
  - Mechanism: The molten salt dissolves the protective oxide scale (e.g., Cr_2O_3, Al_2O_3), exposing the bare metal and accelerating oxidation.
  - Common environments: Turbine blades in high-temperature zones where NaCl and sulfur compounds (SO_3) form Na_2SO_4.
- **Type II Hot Corrosion—Low-Temperature Hot Corrosion**
  - Temperature range: ~650 to 750 °C.
  - Cause: Arises from the presence of sodium and sulfur compounds at lower temperatures before complete oxide layer formation.
  - Mechanism: Sulfur reacts with protective oxides or metals, forming sulfides, which are less stable, porous, and easily spalled.
  - Common environments: Gas turbines during start-up/shutdown or near cooler regions with incomplete combustion.

Chapter 16
Corrosion Testing and Failure Analysis

Corrosion testing is essential for quantifying the corrosion rate of metals and alloys in various environments. It serves multiple purposes, including:

- Evaluating and selecting suitable materials for specific environmental and operational conditions.
- Monitoring and controlling corrosion in engineering systems.
- Advancing research on corrosion mechanisms.

Corrosion testing typically falls into three categories:

1. Laboratory Testing: Conducted in controlled, simulated environments using a limited number of specimens.
2. Pilot Plant Testing: Performed on a larger scale with numerous samples or coupons to replicate more realistic service conditions.
3. Field or Service Testing: Involves exposing materials to actual service environments to assess real-world performance.

Corrosion tests are generally classified into two main types:

- Exposure Tests: Standardized methods (e.g., ASTM or NACE) used to expose materials to defined environmental conditions.
- Electrochemical Tests: Techniques that evaluate corrosion behavior by measuring electrochemical parameters.

One widely used electrochemical technique is Tafel extrapolation (illustrated in Fig. 16.1). In this method, the sample is polarized both anodically and cathodically relative to its open-circuit (corrosion) potential. By analyzing the linear portions of the resulting potential vs. log current density curve (Tafel regions), the corrosion current density (icorr) and corrosion potential (Ecorr) can be determined. These parameters provide a basis for calculating the corrosion rate.

M. M. Ghatus, *Fundamentals of Corrosion Science and Engineering*,
https://doi.org/10.1007/978-3-032-13138-6_16

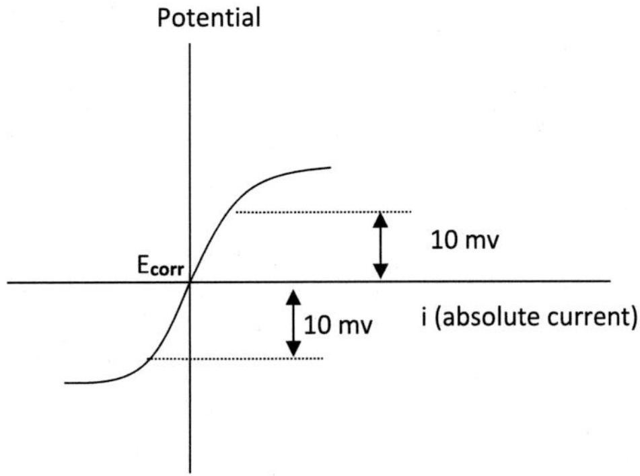

Fig. 16.1 Linear polarization

16.1 Corrosion Rate Testing

- Thus, the corrosion rate can be calculated as follows:

$$\text{Anodic overvoltage } \eta_a = \beta_a \log\left\{\frac{i_a}{i_o}\right\}$$

$$\text{Cathodic overvoltage } \eta_C = \beta_c \log\left\{\frac{i_c}{i_o}\right\}$$

Nenest's equation is:

$$E_{Cath} = E_0 + \frac{RT}{F} * 2.303 \log\left[H^+\right]$$

where:

E_0 of hydrogen is equal to zero and substitutes the values of R, T (298) K, and F, so

$$E_{Cath} = -0.059\text{pH}$$

At E_{corr}, the i_c can be replaced by i_{corr}

$$E_{corr} - E_{Cath} = \beta_C \log\frac{i_{corr}}{i_0}$$

$$E_{corr} + 0.059\text{pH} = \beta_C \log\frac{i_{corr}}{i_0}$$

The last equation finds i_{corr}. Therefore, the corrosion rate can be calculated, as mentioned earlier.

16.2 Linear Polarization

The basic principles of linear polarization are as follows:

Polarize anodically as well as cathodically within a very narrow band, such as 10 mV, the slope of the straight line (potential versus the absolute current density) gives the indication of corrosion rate $\Delta E/\Delta i = i_{corrosion}$ as shown in Fig. 16.1.

Butter-Volumer

$$i_{appl} = i_0 \left\{ \exp\left(\frac{\propto nF}{RT} \eta \right) - \exp\left(-\frac{(1-\propto) nF}{RT} \eta \right) \right\} = i_0 \left(\frac{nF}{RT} \eta \right)$$

→ (linear part in Fig. 16.2 in case of a single electrode system).

$$i_{appl} = i_a - |i_c| \quad \text{(Anodic polarization)}$$

$$i_{appl} = |i_c| - i_a \quad \text{(Cathodic polarization)}$$

$$\eta_a = \left\{ \frac{2.303RT}{+\propto nF} \right\} \log \left\{ \frac{i_a}{i_o} \right\}$$

$$\eta_c = \left\{ \frac{2.303RT}{(i-\propto) nF} \right\} \log \left\{ \frac{i_c}{i_o} \right\}$$

$$\beta_a = \frac{2.303RT}{\propto nF}$$

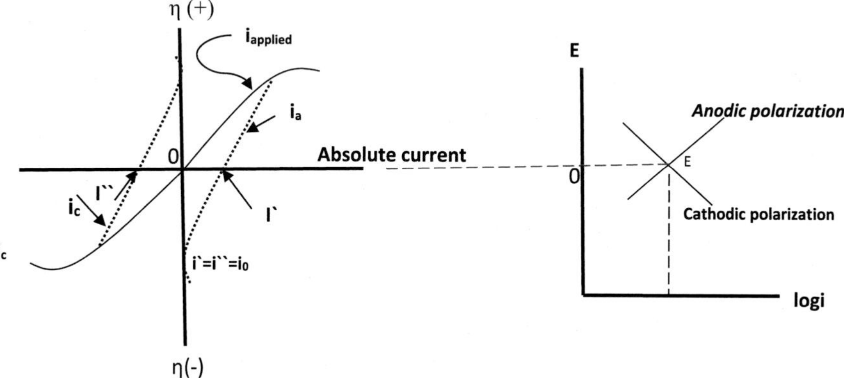

Fig. 16.2 Linear polarization single electrode system

$$\beta_c = \frac{2.303RT}{(1-\propto)nF}$$

Where β_a and β_c are the anodic and cathodic reaction slopes, respectively.

R is the gas constant, T is the temperature, n is the number of electrons, and F is the faraday's constant. Figure 16.2 represents only a single electrode system.

Nevertheless, in the case of corrosion, there are two: one cathodic and one anodic, but overall, the same basic principles of a single electrode system.

In Fig. 16.3, the dimensions of the triangle ABC are shown; the slopes of the anodic and cathodic reactions are indicated as S_a and S_c.

$$S_c = \left(\frac{dE}{di}\right)_{cathodic} = \frac{\Delta E}{\Delta i - x}$$

$$S_a = \left(\frac{dE}{di}\right)_{anodic} = \frac{\Delta E}{x}$$

$$S_c = \frac{1}{\frac{\Delta i}{\Delta E} - \frac{x}{\Delta E}} = \frac{1}{\frac{\Delta i}{\Delta E} - \frac{1}{S_a}}$$

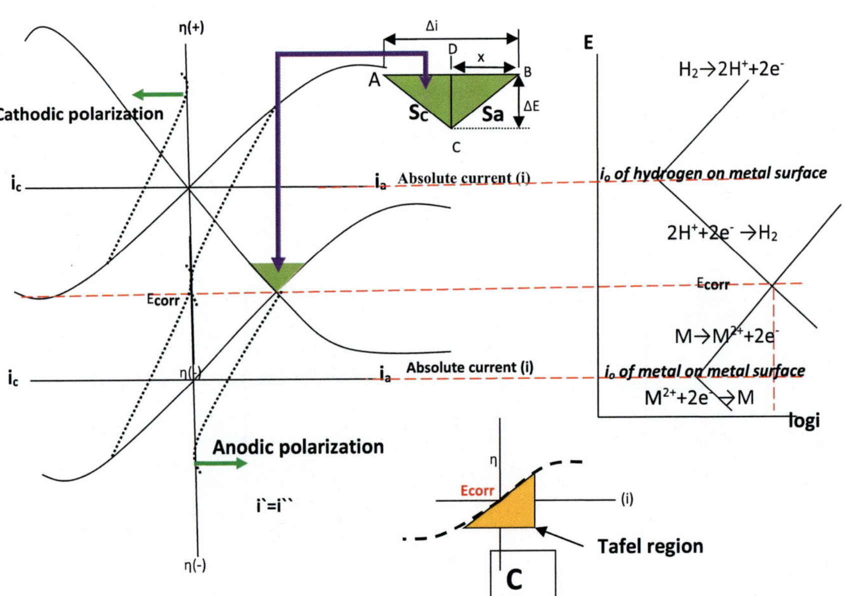

Fig. 16.3 Full linear polarization

$$\frac{\Delta i}{\Delta E} = \frac{1}{S_c} + \frac{1}{S_a} **$$

$$i_a = i_0 \exp\left(\frac{2.303}{\beta_a}\eta_a\right)$$

$$\beta_a = \frac{2.303RT}{\alpha nF}$$

$$\eta_a = \frac{\beta_a}{2.303}\ln\left(\frac{i_a}{i_0}\right)$$

$$\eta_a = E - E_{corr}$$

$$\eta_a = \frac{\beta_a}{2.303}\ln\frac{i_a}{i_0^{Metal}}$$

$$E - E_{corr} = \frac{\beta_a}{2.303}\ln\frac{i_{corr}}{i_0^{Metal}}$$

$$\left(\frac{dE}{di}\right)_{i_{corr}} = \frac{\beta_a}{2.303 i_{corr}} *$$

$$\left(\frac{dE}{di}\right)_{i_{corr}} = S_a$$

$$\left(\frac{dE}{di}\right)_{i_{corr}} = \frac{\beta_c}{2.303 i_{corr}} = S_c *$$

$$\eta_c = \frac{\beta_c}{2.303}\ln\left(\frac{i_{corr}}{i_0^{H_2}}\right)$$

Substitute the values of S_c & S_a (* equations) in double ** equation, so

$$\frac{\Delta i}{\Delta E} = 2.303 i_{corr}\left(\frac{1}{\beta_c} + \frac{1}{\beta_a}\right)$$

$$\frac{\Delta i}{\Delta E} = 2.303 i_{corr}\left(\frac{\beta_a\beta_c}{\beta_c + \beta_a}\right)$$

$$\frac{\Delta E}{\Delta i} = \frac{1}{2.303 i_{corr}} \left(\frac{\beta_a + \beta_c}{\beta_c \beta_a} \right)$$

i_{corr} can be calculated $\Delta E/\Delta i$ from the Figure below (76 C).

The biggest advantage of linear polarization over Tafel's polarization is that it is performed at very low voltage, which reduces the resistance polarization or minimizes the effect of concentration polarization.

Chapter 17
Failure Analysis

Failure analysis is essential for understanding the susceptibility of metals and alloys to corrosion-induced cracking in specific environments. Several test methods are employed to evaluate this behavior:

1. Stress Corrosion Cracking (SCC) Testing: To assess a material's tendency toward SCC, a notched or pre-cracked specimen is immersed in a designated corrosive solution while subjected to a sustained or applied tensile stress. The extent of cracking indicates the susceptibility to SCC (see Fig. 17.1).
2. Slow Strain Rate Testing (SSRT): This method evaluates a material's vulnerability to SCC or hydrogen embrittlement by applying a continuously increasing tensile strain at a slow rate in a corrosive medium.

Failure analysis proceeds through a structured sequence of stages that enable the identification of root causes and the recommendation of preventive or corrective actions:

Stage 1: Preliminary Examination

- Visual inspection of the failed component, often aided by magnification.
- Review of operational records, specifications, and service conditions.
- Photographic documentation of the failure site and affected areas.

Stage 2: Sample Collection

- Retrieval of representative samples from failed regions, including the base material and corrosion products (e.g., rust).

Stage 3: Laboratory Evaluation

- Detailed metallurgical analysis of the samples, including microstructural examination and chemical analysis of corrosion products.

Upon completion of these stages, the analyst integrates all findings to diagnose the failure mechanism and recommend suitable mitigation strategies.

M. M. Ghatus, *Fundamentals of Corrosion Science and Engineering*,
https://doi.org/10.1007/978-3-032-13138-6_17

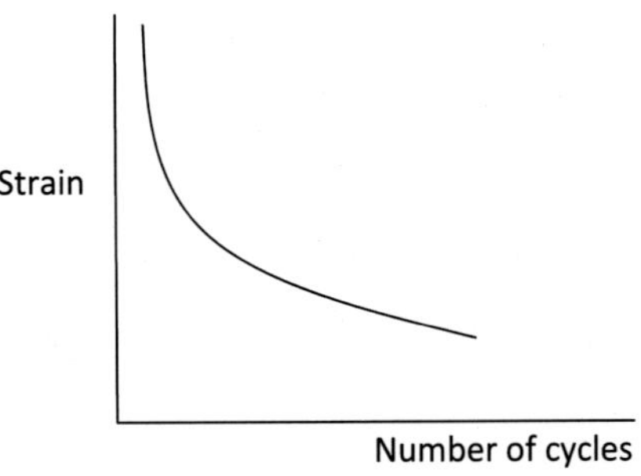

Fig. 17.1 Strain and number of cycles

Chapter 18
Corrosion Behavior of Composite, Polymer, and Ceramic

18.1 Composite

Composite corrosion refers to the corrosion behavior in materials composed of two or more distinct constituents that remain physically identifiable within the structure. As illustrated in Fig. 18.1, a composite typically integrates multiple materials to enhance the overall performance of the system.

For example, consider a laminar composite made from two materials, designated as Material (I) and Material (II), with respective elastic moduli E_1 and E_2. The resulting composite exhibits an intermediate elastic modulus Ec, such that:

$$E_1 < Ec < E_2$$

This arrangement provides a balance of mechanical properties from the individual components. However, when exposed to corrosive environments, each constituent may behave differently due to variations in their electrochemical reactivity, potentially leading to galvanic or localized corrosion within the composite system. Proper material selection and protective strategies are essential to mitigate such risks.

18.1.1 Corrosion in Reinforced Concrete Structures

As shown in Fig. 18.2, a reinforced concrete structure is a composite material comprising concrete and embedded reinforcing steel. Concrete itself consists of cement, aggregates (stones), and various chemical additives, including calcium hydroxide, sodium hydroxide, and potassium hydroxide. These alkaline compounds elevate the pH of the environment surrounding the steel reinforcement to a highly alkaline range (pH 9.5–14).

M. M. Ghatus, *Fundamentals of Corrosion Science and Engineering*,
https://doi.org/10.1007/978-3-032-13138-6_18

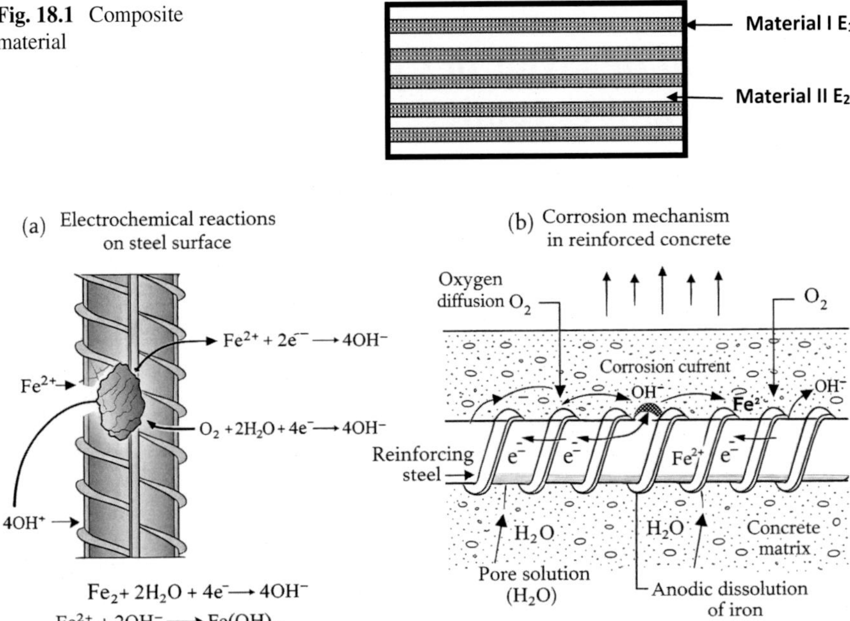

Fig. 18.1 Composite material

Fig. 18.2 Concrete corrosion

According to the Pourbaix diagram (Fig. 18.3), steel in this pH range is immune or passivated, meaning it is protected against corrosion. However, this protective condition can deteriorate due to various environmental factors.

Cracks, micro-porosity, or permeability in the concrete matrix allow aggressive species such as oxygen, water, and chloride ions to penetrate and reach the steel surface. The ingress of these agents gradually lowers the local pH to between 7 and 8, moving the steel from the passive to the active corrosion zone. This reduction in pH can initiate localized corrosion mechanisms such as pitting and crevice corrosion.

When corrosion products like iron oxides accumulate, they expand in volume. Since concrete is brittle, this expansion can lead to internal cracking, commonly called concrete cancer. Additionally, thermal expansion and contraction of the reinforcing steel during seasonal temperature variations may lead to debonding from the surrounding concrete, further increasing corrosion risk.

Another contributing factor is stray current corrosion, which occurs when unintended electrical currents pass through the concrete and exit the steel reinforcement, accelerating material loss at exit points.

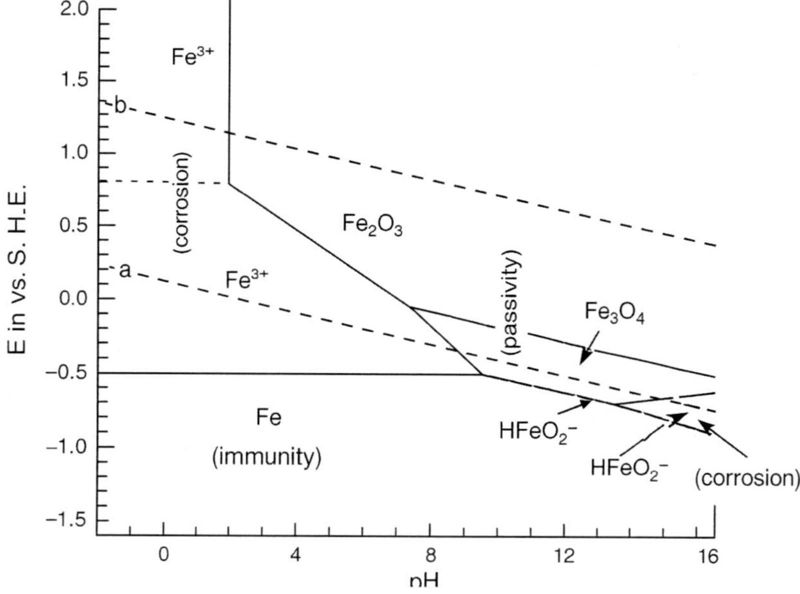

Fig. 18.3 Pourbaix diagram of iron

18.1.2 Preventive Measures Against Reinforced Concrete Corrosion

1. Maintain a dry environment to limit moisture ingress.
2. Ensure a dry atmospheric condition, particularly in storage or structural exposure.
3. Apply epoxy coatings on reinforcement bars to create a barrier against corrosive agents.
4. Use improved materials, such as stainless steel reinforcement, for enhanced corrosion resistance.
5. Incorporate fly ash in the concrete mix to reduce permeability and slow chloride diffusion.

18.2 Corrosion of Polymer Materials

Polymers are macromolecular materials composed of repeating structural units known as monomers. These monomers are typically hydrocarbon-based compounds. A common example is polyethylene, which is synthesized from the ethylene monomer (C_2H_4) through a polymerization reaction. This transformation involves breaking the double bond in ethylene and forming long chains of saturated hydrocarbons.

Unlike metals, polymers do not undergo electrochemical corrosion. Instead, their degradation often referred to as polymer corrosion, results from chemical, thermal, mechanical, photochemical, or biological attacks. The key mechanisms of polymer degradation include:

- Oxidative degradation: Reaction with oxygen, often accelerated by heat or UV radiation, leading to chain scission and embrittlement.
- Hydrolysis: Reaction with water, particularly in polymers containing ester, amide, or carbonate linkages.
- Solvent attack: Swelling, dissolution, or plasticization of the polymer in contact with certain organic or inorganic solvents.
- UV radiation: Breakdown of polymer chains due to exposure to ultraviolet light (photodegradation).
- Microbial degradation: Attack by bacteria or fungi, especially in biodegradable or natural polymers.

While polymers are generally resistant to many corrosive environments, their long-term performance can be significantly affected by the combined effects of environmental exposure and mechanical stress.

18.3 Degradation of Polymers

Polymer degradation refers to the deterioration of a polymer's properties due to physical, chemical, or environmental influences. This process may lead to changes in appearance, mechanical strength, flexibility, or overall performance. The main mechanisms include:

1. Swelling and Distortion

   Water or organic solvents can penetrate the polymer matrix, leading to swelling. Water molecules, for example, insert themselves between polymer chains, weakening secondary bonds (e.g., Van der Waals or hydrogen bonding). This swelling often results in softening, deformation, and eventual structural weakening.

   - *Example:* Hydrocarbon-based rubbers absorb hydrocarbon liquids like gasoline, causing excessive swelling.
   - Conversely, many polymers exhibit good resistance to strong acids such as hydrofluoric acid (HF).

2. Bond Rupture by Radiation or Chemical Attack

   - Radiation-induced degradation: High-energy radiation such as X-rays, gamma rays, or ultraviolet (UV) light can break covalent bonds in the polymer backbone. This may result in brittleness, discoloration, cracking, or chalking.

  - *Example:* UV-induced cracking in automotive dashboards or outdoor camping plastics.

- Chemical degradation: Oxidizing agents like ozone (O_3) or oxygen (O_2) can react with unsaturated or vulcanized polymers, breaking bonds and reducing molecular weight.
- Thermal degradation: Elevated temperatures may induce chemical reactions that generate gaseous by-products or degrade functional groups.
  - *Example:* Heating polyvinyl chloride (PVC) can cause dehydrochlorination, releasing hydrogen chloride (HCl) gas and resulting in discoloration and brittleness.

3. Weathering

   Environmental exposure involving sunlight, moisture, oxygen, and ozone leads to weathering, a combined degradation mechanism. UV light initiates photodegradation, while moisture and pollutants accelerate the loss of mechanical integrity.

   - *Example:* Fading, cracking, or loss of gloss in outdoor plastic components.

4. Other Degradation Mechanisms

   - Oxidation: Autoxidation of susceptible polymers results in chain scission and embrittlement.
   - Ultrasound exposure: High-frequency vibrations may induce localized heating and mechanical stress, promoting micro-cracks or void formation.

18.4 Degradation of Ceramics

Ceramics are generally known for their excellent corrosion resistance, making them suitable for use in aggressive chemical environments. However, they also possess certain limitations, including brittleness and poor resistance to thermal shock.

Key characteristics of ceramics include:

- High melting temperatures, which allow them to withstand elevated thermal conditions.
- Chemical stability, especially in acidic and aqueous environments.
- Inertness, making them suitable for handling strong acids and resisting moisture penetration.

Examples and Applications

- Glass (a non-crystalline ceramic) is widely used in chemical industries for acid handling due to its strong resistance to corrosive fluids and moisture.
- Aluminosilicate ceramics exhibit excellent resistance to water but are less effective against acidic attack.

- Borosilicate glass, by contrast, offers superior resistance to both acids and alkalis, making it ideal for laboratory glassware and industrial piping systems.
- Acid-resistant bricks, made from specialized ceramic materials, are often used to line storage tanks and process vessels, especially where high acid concentrations are present.

Bibliography

1. Fontana, M. G. (1987). *Corrosion engineering* (3rd ed.). McGraw-Hill.
2. Evans, U. R. (1960). *Corrosion and oxidation of metals*. Edward Arnold.
3. Uhlig, H. H., & Revie, R. W. (1985). *Corrosion and corrosion control* (3rd ed.). John Wiley & Sons.
4. Shreir, L. L. (1976). *Corrosion* (Vols. 1–2). Newnes-Butterworths.
5. Jones, D. A. (1996). *Principles and prevention of corrosion* (2nd ed.). Prentice Hall.
6. Rozenfeld, I. L. (1981). *Corrosion inhibitors*. McGraw-Hill.
7. Pourbaix, M. (1974). *Atlas of electrochemical equilibria in aqueous solutions* (2nd ed.). NACE International.
8. McCafferty, E. (2010). *Introduction to corrosion science*. Springer.
9. ASM International. (2003). *ASM handbook: Volume 13A—Corrosion: Fundamentals, testing, and protection*. ASM International.
10. Milinković, M. (1970). *Korozija i Zastita*. Tehnička Knjiga.
11. Vučković, I. (1966). *Korozija*. Novinsko Izdavačko Preduzeće Tehnička Knjiga.
12. Moore, W. J. (1972). *Physical chemistry*. Longman.
13. Nalco Chemical Company. (1972). *The Nalco water handbook*. McGraw-Hill.
14. Degremont. (1979). *Water treatment handbook*. Halsted Press.
15. Schweitzer, P. A. (2006). *Fundamentals of metallic corrosion: Atmospheric and media corrosion of metals*. CRC Press.
16. Callister, W. D., & Rethwisch, D. G. (2014). *Materials science and engineering: An introduction* (9th ed.). Wiley.
17. ASTM International. ASTM G1–03(2017): Standard practice for preparing, cleaning, and evaluating corrosion test specimens.
18. NACE International (Ed.). (2005). *Corrosion basics: An introduction* (2nd ed.). NACE International.

M. M. Ghatus, *Fundamentals of Corrosion Science and Engineering*,
https://doi.org/10.1007/978-3-032-13138-6